Agriculture and Climate Change: An Agenda for Negotiation in Copenhagen

2020 VISION FOR FOOD, AGRICULTURE, AND THE ENVIRONMENT

Focus 16 • Introduction • May 2009

Agriculture and climate change are inextricably linked.

Agriculture is part of the climate change problem, contributing about 13.5 percent of annual greenhouse gas (GHG) emissions (with forestry contributing an additional 19 percent), compared with 13.1 percent from transportation. Agriculture is, however, also part of the solution, offering promising opportunities for mitigating GHG emissions through carbon sequestration, soil and land use management, and biomass production.

Climate change threatens agricultural production through higher and more variable temperatures, changes in precipitation patterns, and increased occurrences of extreme events such as droughts and floods. And if agriculture is not included, or not well included, in the international climate change negotiations leading up to the 15th Conference of Parties (COP15) of the UN Framework Convention on Climate Change in Copenhagen in December 2009, resulting climate change policies could threaten poor farming communities and smallholders in many developing countries. The policies could also impede the ability of smallholders to partake in new economic opportunities that might arise from the negotiations.

Therefore, agriculture must be on the Copenhagen agenda. Indeed, it must be on the agenda of negotiators well before COP15. Essentially, three avenues must be pursued:

(1) **Investments.** There must be explicit inclusion of agriculture-related investments, especially as part of a Global Climate Change Fund.

(2) **Incentives.** There must be a deliberate focus on introducing incentives to reduce emissions and support technological change.

(3) **Information.** There must be a solid commitment to establishing comprehensive information and monitoring services in soil and land use management for verification purposes.

For more comprehensive perspectives on these issues, IFPRI's 2020 Vision Initiative approached leading experts around the world to share their views on the key negotiating outcomes that must be pursued now in order to effectively put agriculture on the climate change agenda. We are grateful to Gerald Nelson for conceptualizing and editing this collection of policy briefs, to the contributors for their analysis and insights, and to the reviewers for their constructive comments. We hope that the findings and suggested negotiating outcomes presented here will contribute to a decisive and meaningful inclusion of agriculture in the international climate change negotiations leading up to Copenhagen in December 2009.

Joachim von Braun
Director General

Rajul Pandya-Lorch
Head, 2020 Vision Initiative

The International Food Policy Research Institute (IFPRI) is one of several international research centers supported by the Consultative Group on International Agricultural Research (CGIAR).

"2020 Vision for Food, Agriculture, and the Environment" is an initiative of IFPRI to develop a shared vision and consensus for action on how to meet future world food needs while reducing poverty and protecting the environment.

IFPRI gratefully acknowledges the generous unrestricted funding from Australia, Canada, China, Finland, France, Germany, India, Ireland, Italy, Japan, Netherlands, Norway, South Africa, Sweden, Switzerland, United Kingdom, United States, and World Bank.

Agriculture and Climate Change: An Agenda for Negotiation in Copenhagen

Overview

Gerald C. Nelson

2020 Vision for Food, Agriculture, and the Environment

Focus 16 • Brief 1 • May 2009

If fundamental climate change mitigation and adaptation goals are to be met, international climate negotiations must include agriculture. Agriculture and climate change are linked in important ways, and this brief focuses on three: (1) climate change will have large effects on agriculture, but precisely where and how much are uncertain, (2) agriculture can help mitigate climate change, and (3) poor farmers will need help adapting to climate change. As negotiations get underway in advance of the meeting of the 15th Conference of Parties of the UN Framework Convention on Climate Change (UNFCCC) in Copenhagen in December 2009, this brief suggests negotiating outcomes for both adaptation and mitigation that will support climate change goals while enhancing the well-being of people who manage and depend on agriculture, especially in the developing world.

Goal

Put agriculture on the agenda of the UNFCCC negotiations in Copenhagen

Climate change will affect agriculture, but it is uncertain where and how much

Climate change will have dramatic consequences for agriculture. Water sources will become more variable, droughts and floods will stress agricultural systems, some coastal food-producing areas will be inundated by the seas, and food production will fall in some places in the interior. Developing economies and the poorest of the poor likely will be hardest hit. Overall, however, substantial uncertainty remains about where the effects will be greatest.

Agricultural outcomes are determined by complex interactions among people, policies, and nature. Crops and animals are affected by changes in temperature and precipitation, but they are also influenced by human investments such as irrigation systems, transportation infrastructure, and animal shelters. Given the uncertainties about where climate change will take place and how farmers will respond, much is still unknown about the effects of climate change on agricultural production, consumption, and human well-being, making it difficult to move forward on policies to combat the effects of climate change.

Suggested negotiating outcome: Fund research on the interactions between climate change and agriculture

Research that improves understanding and predictions of the interactions between climate change and agriculture should be funded. Climate change assessment tools are needed that are more geographically precise, that are more useful for agricultural policy and program review and scenario assessment, that more explicitly incorporate the biophysical constraints that affect agricultural productivity, and that better integrate biophysical and socioeconomic scenarios.

Cost-effective ways to help poor farmers adapt to climate change are needed

Even with the best efforts to mitigate greenhouse gases (GHGs), it is inevitable that poor farmers will be affected. The goal is to find and fund the most cost-effective ways to help the poor adapt to the changes, a daunting task because of uncertainty about the magnitude of possible changes, their geographic distribution, and the long lead times needed to implement adaptation efforts.

Suggested negotiating outcome: Allow funding mechanisms that recognize the connection between pro-poor development policies for sustainable growth and sound climate change policies

A pro-growth, pro-poor development agenda that supports agricultural sustainability also contributes to climate change adaptation. Adaptation is easier when individuals have more resources at their command and operate in an economic environment with the flexibility to respond quickly to changes. If, as seems likely, the effects of climate change will fall disproportionately on poor farmers, a policy environment that enhances opportunities for smallholders will also be good for climate change adaptation. Such an environment would include more investment in agricultural research and extension, rural infrastructure, and access to markets for small farmers. Funding should support these kinds of policy changes and investments in institution building.

Suggested negotiating outcome: Allow funding mechanisms that recognize and support synergies between adaptation and mitigation

Many changes to management systems that make them more resilient to climate change also increase carbon sequestration. Conservation tillage increases soil water retention in the face of drought while also sequestering carbon below ground. Small-scale irrigation facilities not only conserve water in the face of greater variability, but also increase crop productivity and soil carbon. Agroforestry systems increase above- and below-ground carbon storage while also increasing water storage below ground, even in the face of extreme climate events. Properly managed rangelands can cope better with drought and sequester significant amounts of carbon. Project- and program-based funding schemes that support adaptation should also be able to draw on mitigation resources.

Suggested negotiating outcome: Provide funds for agricultural science and technology

Even without climate change, greater investments in agricultural science and technology are needed to meet the demands of a world population expected to reach 9 billion by 2050. Many of these people will live in the developing world, have higher incomes, and desire a more diverse diet. Agriculture science- and technology-based solutions are essential to meet those demands.

Climate change places new and more challenging demand on agricultural productivity. It is urgent to pursue crop and livestock research, including biotechnology, to help overcome stresses related to climate change such as heat, drought, and novel pathogens. Crops and livestock are needed that respond reasonably well in a range of production environments rather than extremely well in a narrow set of climate conditions. Research is also needed on how dietary changes in food animals, including pasture improvements, can reduce methane emissions.

One of the key lessons of the Green Revolution is that improved agricultural productivity, even if not targeted to the poorest of the poor, can be a powerful mechanism for alleviating poverty indirectly by creating jobs and lowering food prices. Productivity enhancements that increase farmers' resilience in the face of climate change pressures will likely have similar poverty-reducing effects.

Suggested negotiating outcome: Provide funds for infrastructure and institutional innovations

Improvements in water productivity are critical, and climate change, by making rainfall more variable and changing its spatial distribution, will exacerbate the need for better water harvesting, storage, and management. Equally important is supporting innovative institutional mechanisms that give agricultural water users incentives to conserve.

Investments in rural infrastructure, both physical (such as roads, market buildings, and storage facilities) and institutional (such as extension programs, credit and input markets, and reduced barriers to internal trade) are needed to enhance the resilience of agriculture in the face of the uncertainties of climate change.

Suggested negotiating outcome: Provide funds for data collection on the local context of agriculture

Agriculture is an intensely local activity. Crop and livestock productivity, market access, and the effects of climate are all extremely location-specific. Yet national and global efforts to collect and disseminate data on the spatial nature of agriculture, especially over time, are poorly developed. Countries have reduced funding for national statistical programs, and remote sensing systems are still inadequate to the task of monitoring global change. Understanding agriculture-climate interactions well enough to support adaptation and mitigation activities based on land use requires major improvements in data collection and provision.

Agriculture can help mitigate GHG emissions

Today, agriculture contributes about 14 percent of annual GHG emissions, and land use change including forest loss contributes another 19 percent. The relative contributions differ dramatically by region. The developing world accounts for about 50 percent of agricultural emissions and 80 percent of land use change and forestry emissions.

The formal inclusion of REDD (Reducing Emissions from Deforestation and forest Degradation) in the current negotiations is a result of a new appreciation of the importance of this source of GHGs and initial findings of low-cost opportunities to reduce them. Significant challenges remain, however. What are the best approaches to dissuade poor people from cutting down trees and converting other lands to unsustainable agricultural practices and to instead encourage them to adopt technologies and management strategies that mitigate carbon, methane, and nitrous oxide emissions? The tasks ahead include identifying and supporting the most appropriate approaches in farmers' fields and monitoring their implementation.

Suggested negotiating outcome: Fund cost-effective mitigation in agriculture and research on promising technologies and management systems

Agriculture has huge potential to cost-effectively mitigate GHGs through changes in agricultural technologies and management practices. Changing crop mixes to include more plants that are perennial or have deep root systems increases the amount of carbon stored in the soil. Cultivation systems that leave residues and reduce tillage, especially deep tillage, encourage the buildup of soil carbon. Shifting land use from annual crops to perennial crops, pasture, and agroforestry increases both above- and below-ground carbon stocks. Changes in crop genetics and the management of irrigation, fertilizer use, and soils can reduce both nitrous oxide and methane emissions. Changes in livestock species and improved feeding practices can also cut methane emissions. Mitigation funding programs arising from the negotiations should thus include agriculture.

Suggested negotiating outcome: Fund low-cost systems for monitoring agricultural mitigation

It is much easier to monitor 1,500 U.S. coal-fired power plants than several million smallholder farmers who rely on field, pasture, and forest for their livelihoods. Nonetheless, promising technologies exist for reducing the costs of tracking the performance of agricultural mitigation programs. For example, microsatellites can be used for frequent, high-resolution land cover imaging, inexpensive standardized methods are available to test soil carbon, and simple assessment methods can adequately quantify the effects of management technologies on methane and nitrous oxide emissions. These monitoring technologies and others require funding.

Suggested negotiating outcome: Allow innovative payment mechanisms and support for novel institutions for agricultural mitigation

Agricultural production differs qualitatively from other sources of GHGs in that the sources are individually small, geographically dispersed, and often served by inadequate physical and institutional infrastructure. Cost-effective payment mechanisms to encourage agricultural mitigation must reflect these differences. Beyond the traditional schemes developed under the Kyoto Protocol, the negotiating outcome should allow and encourage alternatives that take advantage of these differences, exploiting activities beyond project-specific funding. Examples include land retirement contracts, one-time payments for physical infrastructure investments that have long-term mitigation effects, and payments for institutional innovations that encourage mitigating behavior in common property resources.

Concluding Remarks

Agricultural activities around the world are responsible for almost 15 percent of annual GHG emissions. They could be an important sink for emissions from other sectors and are likely to be altered dramatically by climate change. Agriculture also provides a living for more than half of the world's poorest people. The ongoing negotiations to address climate change provide a unique opportunity to combine low-cost mitigation and essential adaptation outcomes with poverty reduction. ■

Gerald C. Nelson *(g.nelson@cgiar.org) is a senior research fellow at the International Food Policy Research Institute (IFPRI).*

International Food Policy Research Institute
2033 K Street, NW • Washington, DC 20006-1002 • USA
Phone: +1-202-862-5600 • Skype: ifprihomeoffice • Fax: +1-202-467-4439 • Email: ifpri@cgiar.org
IFPRI® **www.ifpri.org**

Agriculture and Climate Change: An Agenda for Negotiation in Copenhagen

Agricultural Science and Technology Needs for Climate Change Adaptation and Mitigation

Rudy Rabbinge

2020 Vision for Food, Agriculture, and the Environment

Focus 16 • Brief 2 • May 2009

Higher temperatures, more variable precipitation, and changes in the frequency and severity of extreme climate events will have significant consequences for food production and food security. However, the frequency of heat stress, drought, and flooding are also expected to increase, even though they cannot be modeled satisfactorily with current climate models. They will undoubtedly have adverse effects on crops and agricultural productivity over and above the effects due to changes in mean variables alone. The impacts of climate change on agriculture are likely to be regionally distinct and highly heterogeneous spatially, requiring sophisticated understanding of causes and effects and careful design and dissemination of appropriate responses.

These changes will challenge the livelihoods of farmers, fishers, and forest-dependent people who are already vulnerable and food insecure. Adapting to these changes, while continuing to feed a world of 9 billion people, requires the formation of a global partnership in science, technology development, and dissemination of results to millions of smallholder farmers, bringing together research workers and resource managers from many fields. To take an international approach to climate change, new partnerships must be forged, linking the agricultural research and climate science communities.

The CGIAR

The Consultative Group on International Agricultural Research (CGIAR) is a strategic partnership, whose members support 15 international centers and five major collaborative programs around the world. The CGIAR plans to contribute its broad-based and multidisciplinary experience in developing-country agriculture to global efforts to adapt to and mitigate climate change through research on agriculture and natural resources. Work already underway that is directly applicable to climate change research includes breeding crops for stress tolerance; developing better practices for sustainable crop and environmental management; gauging the vulnerability of agriculture, natural resources, and rural communities; and supporting the development of policies conducive to sustainable agricultural growth.

The CGIAR's Consortium for Spatial Information is taking the initiative with other centers of excellence to create a climate information portal for mapping data. The Climate Change Challenge Program is uniting the expertise of the CGIAR with the Earth System Science Partnership to close critical knowledge gaps on how to deal with trade-offs among food security, livelihood, and environmental goals as climate changes.

Challenges in defining the effects of change

Climate modeling and scenario building are important for the global approach to agricultural research, but three challenges must be addressed to achieve practical results. First, we must understand what the local impacts of climate change are likely to be. Uncertainties are involved in scaling down the global climate model output to the high spatial resolutions needed for effective adaptation work at regional and national levels. Some archipelagic and small island countries most at risk from climate change barely figure in global models. Substantiating the local effects of long-term change requires that long-term research and monitoring is supported at key, agro-ecologically defined regional sites. Second, a significant gap exists between the seasonal information we currently have and that on climate change in the long run (2050 and beyond): information about what is likely over the next 3 to 20 years is largely missing. This presents a critical problem, as this time scale is vital for vulnerability assessment, agricultural planning, and political negotiation. Third, convincingly communicating the results from modeling scenarios to decision-makers, including farmers and policymakers, will be one of the most significant challenges. Scenarios integrating possible socioeconomic (and climate) futures will therefore be central to exploring and communicating adaptation and mitigation approaches. There must be a long-term approach to building knowledge and capacity at the local scale for effective responses to occur.

Challenges to crop agriculture

The current climate change scenarios demand adaptation to temperature increases, changing amounts of available water, climatic instability and increased frequency of extreme weather events, and rises in sea level and saline intrusion in the coastal zones. Thus future crop farming and food systems will have to be better adapted to a range of abiotic stresses (such as heat or salinity) and biotic stresses (such as pests) to cope with the consequences of a progressively changing climate. In response, the CGIAR is working on gene discovery and improving plant tolerance for heat, drought, and submergence. This work should be expanded to consider the basic energy and water efficiency of plants (improving their photosynthetic capacity and reducing evapotranspiration).

Crop germplasm improvement, natural resource management, and inclusion of enhanced agrobiodiversity have a proven track record of decreasing susceptibility to individual stresses. Breeding and marker-assisted selection are important mechanisms for introducing improved characteristics and achieving yield improvements for most crops. Defining future targeted farming systems and their environments could allow breeding and management programs to be matched with georeferenced data on crop germplasm collections. This would allow the identification of crops and cultivars best suited to predicted conditions, based on the agro-ecological parameters of their places of origin. Improved water-management approaches, with conservation agriculture, are likely to be central to adaptation strategies in both irrigated and dryland agriculture. Work on feed plants (for livestock and aquaculture) should be incorporated into this research approach. However, technical innovations will not be sufficient on their own. Strengthening the adaptive capacities of farmers and other land users requires a variety of strategies ranging from diversification of production systems to improved institutional settings. It is crucial to add value to current investment in agronomic crop management and germplasm improvements by integrating new results and best practices from these fields into adaptation options proposed in the policy domain. There may well be major land use changes, and research will be needed to identify and assess options to support the transitions this will impose on farmers and other actors within the food system.

Challenges to livestock agriculture

Livestock are a critical component of agriculture, particularly for the income and nutrition of the poor in developing countries. However, the magnitude of the changes that are likely to befall livestock systems is a relatively neglected research area. Little is known about the interactions of climate and increasing climate variability with other drivers of change in livestock systems and in broader development trends. While opportunities may exist for some households to take advantage of more conducive rangeland and cropping conditions, for example, the changes projected will pose very serious problems for many other households. Furthermore, ruminant livestock themselves have important impacts on climate, through the emission of methane and through the land-use change that may be brought about by livestock keepers. Nevertheless, meeting anticipated demand for meat and milk and other necessary livestock products in the coming decades will require attention to the supply of livestock feeds. Climate change sharpens the edge of the production dilemma among human food, animal feed, and (potentially) energy on a finite amount of land. The issue of temperature and other abiotic stresses will have to be as carefully addressed in feed plants as in human crops. Critically, altered climate regimes will alter the ranges of insect pests and vectors; a major risk of climate change is that it will change or extend the range of current diseases or, through unknown effects, create the conditions for the spread of new diseases to the livestock population. Human health would also be threatened by an increase in zoonotic diseases. Since the impacts of climate change on livestock disease may be extremely complex, integrated approaches must go well beyond climate and risk mapping and will require epidemiological reconnaissance, new diagnostic reagents, adapted or new livestock genotypes, and new veterinary and public health management services.

Challenges to forests and forestry

Major recent fora on forestry have concluded that integrated approaches to adaptive forest management are a central component of the global response to climate change. Within global approaches, there is the opportunity to both reduce forest destruction and potentially to sequester carbon (atmospheric CO_2) as a climate change mitigation measure. Test cases for the payment for environmental services approach are being tried in the forestry sector (for example, Reduced Emissions from Deforestation (REDD) payment schemes). There should be continuous scientific, economic, and social evaluation of Payments for Environmental Services (PES) schemes so that their true value to the environment and to the lives of the poor are put on an evidence-based footing.

Challenges to fisheries

Fisheries are key natural resources ensuring food security for large numbers of people. Successful fisheries depend upon coherent marine and freshwater ecosystems, which are at risk of disruption by climate change. As temperatures change, fisheries are likely to gradually be displaced or migratory patterns may become erratic, affecting fish supplies for both human consumption and aquaculture and livestock feeds. There could be long-term effects on coral reefs (which are very susceptible to small changes in temperature). The rise in coastal sea levels could disrupt livelihoods and cause salt water intrusion into agricultural land. Like livestock industries, aquaculture competes for feed resources (from aquatic or terrestrial sources). A broad set of tasks, linking research assessment and monitoring of fisheries to the design of adaptive measures and appropriate policies, must be addressed to sustain poor communities through the expected changes. Aquaculture will require that particular attention be given to the breeding of robust genotypes and the design of sustainable feed resource policies.

These tasks will require cooperative approaches among research providers across the fields of agricultural and climate change science. New collaborative arrangements will need to be implemented, with each organization playing its part according to its comparative advantage.

Suggested negotiating outcomes:

- The world currently has imperfect knowledge of agriculture–climate change interactions. In order to effectively plan and implement adaptation strategies, funds must be made available to increase knowledge in this area, particularly for determining more precise climate change effects on developing-country agriculture, forestry, and fisheries.
- The development of adaptation strategies for developing-country agriculture should be a key element of any adaptation fund. International exchange of information and collaboration among science groups in different sectors should be fostered and supported. Implementation of integrated strategies for adaptation to climate change that affect farmers will require funding for indirect adaptation expenditures, such as improvements to national agricultural research and extension services, as well as international research and project-specific funding.
- Mitigation strategies should primarily address global energy policy. However, some sectors of agriculture are net contributors to global greenhouse gas (GHG) emissions. Investigation into whether there is potential for low-cost effective sequestration of GHGs by agricultural systems should be supported. ■

For Further Reading: Consultative Group on International Agricultural Research (CGIAR), "Global Climate Change: Can Agriculture Cope?" (2009) www.cgiar.org/impact/global/climate.html; CGIAR/Earth System Science Partnership (ESSP), "Climate Change, Agriculture, and Food Security – A Proposal for a CGIAR Challenge Program" (2008) www.cgiar.org/impact/challenge/cccp.html; International Livestock Research Institute (ILRI) "The Livestock-Climate-Poverty Nexus: A Discussion Paper on ILRI Research in Relation to Climate Change" (2008) www.ilri.org/Infoserv/webpub/fulldocs/DiscuPaper11_Climate/Livesto_Climat_Pover_Nexus_DP11.pdf; Center for International Forestry Research (CIFOR) and Centro Agronómico Tropical de Investigación y Enseñanza (CATIE), "Tropical Forests and Climate Change Adaptation" (2008) www.cifor.cgiar.org/trofcca/welcome.asp; E. H. Allison et al., "Vulnerability of National Economies to the Impacts of Climate Change on Fisheries," *Fish and Fisheries* 10 (2009): 173-196, available at www.worldfishcenter.org/wfcms/HQ/article.aspx?ID=223.

***Rudy Rabbinge** (rudy.rabbinge@wur.nl) is chair of the CGIAR's Science Council.*

International Food Policy Research Institute
2033 K Street, NW • Washington, DC 20006-1002 • USA
Phone: +1-202-862-5600 • Skype: ifprihomeoffice • Fax: +1-202-467-4439 • Email: ifpri@cgiar.org
IFPRI® **www.ifpri.org**

AGRICULTURE AND CLIMATE CHANGE: AN AGENDA FOR NEGOTIATION IN COPENHAGEN

Reducing Methane Emissions from Irrigated Rice

REINER WASSMANN, YASUKAZU HOSEN, AND KAY SUMFLETH

2020 VISION FOR FOOD, AGRICULTURE, AND THE ENVIRONMENT

FOCUS 16 • BRIEF 3 • MAY 2009

Rice is grown on more than 140 million hectares worldwide and is the most heavily consumed staple food on earth. Ninety percent of the world's rice is produced and consumed in Asia, and 90 percent of rice land is—at least temporarily—flooded. The unique semiaquatic nature of the rice plant allows it to grow productively in places no other crop could exist, but it is also the reason for its emissions of the major greenhouse gas (GHG), methane.

Methane emissions from rice fields are determined mainly by water regime and organic inputs, but they are also influenced by soil type, weather, tillage management, residues, fertilizers, and rice cultivar. Flooding of the soil is a prerequisite for sustained emissions of methane. Recent assessments of methane emissions from irrigated rice cultivation estimate global emissions for the year 2000 at a level corresponding to 625 million metric tons (mt) of carbon dioxide equivalent (CO_2e).

Midseason drainage (a common irrigation practice adopted in major rice growing regions of China and Japan) and intermittent irrigation (common in northwest India) greatly reduce methane emissions. Similarly, rice environments with an insecure supply of water, namely rainfed rice, have a lower emission potential than irrigated rice. Organic inputs stimulate methane emissions as long as fields remain flooded. Therefore, organic inputs should be applied to aerobic soil in an effort to reduce methane emission. In addition to management factors, methane emissions are also affected by soil parameters and climate.

Accounting for nitrous oxide (N_2O) and CO_2 emissions

Recent studies suggest that rice cultivation is an important anthropogenic source of not only atmospheric methane but also of N_2O. Rice soils that are flooded for long periods of the year tend to accumulate soil organic carbon, even with complete removal of above-ground plant biomass. Significant input of carbon and nitrogen is derived from biological activity in the soil–floodwater system, and conditions are generally more favorable for the formation of conserved soil organic matter. It is currently unknown whether rice systems in the tropics and subtropics truly sequester atmospheric carbon and how soil organic carbon levels will react to a changing climate or new management practices.

Losses of soil organic carbon are of major concern for certain developments in the agricultural sector that are undergoing rapid intensification and diversification of crop land. At the International Rice Research Institute (IRRI), however, 12 years of continuous rice cropping in flooded fields did not cause any significant decline in soil organic carbon. In contrast, the soil organic carbon immediately declined after a shift to a nonflooded system, namely maize. The modification of flooding patterns encompassing one or more dry periods may somehow accelerate decomposition, but—unlike a complete shift to upland systems—the recurring periods of flooding will keep the overall soil organic carbon at a fairly stable level. Thus, we do not include CO_2 emissions in our considerations of mitigation options.

Mitigation options

Changing water management appears to be the most promising mitigation option and is particularly suited to reducing emissions in irrigated rice production. Midseason drainage and intermittent irrigation reduce methane emissions by over 40 percent. Shallow flooding provides additional benefits, including water conservation and increased yields. A recent study estimates large potential for additional methane reductions from Chinese rice paddies through modifications of water-management strategies, even though midseason drying is widely practiced there.

Midseason drainage or reduced water use also creates nearly saturated soil conditions, which may promote N_2O production. There are conflicting reports on the net global warming potential (GWP) of midseason drainage, but there seems to be a growing consensus that this practice decreases the net GWP of paddy fields as long as nitrogen is applied in appropriate doses. According to an empirical model proposed by Yan et al. (2005), midseason drainage generally tends to be an effective option for mitigating net GWP, although 15 to 20 percent of the benefit gained by decreasing methane emission was offset by the increase in N_2O emission.

We can conclude that midseason drainage has a potential to be an effective option to mitigate the net GWP from rice fields, especially when larger amounts of rice straw are returned into the soil. However, there is the risk that N_2O emission offsets the reduction of methane emission when nitrogen fertilizer is applied at a high rate. Therefore, modifications of water regime should be coupled with efficient fertilizer application in order to reduce both GHG emissions and costs (for irrigation water and fertilizers).

The immense variability of environmental factors affecting the 140 million hectares of annually harvested rice fields denies the use of blanket strategies to reduce emissions. Moreover, technological options in rice production have to remain economically viable despite rapid changes in both socioeconomic development and the environment. Two case studies looking at two different countries—India and the Philippines—at vastly different scales, illustrate the mitigation potential of water regime modifications.

Case study: Country-wide mitigation in India

Indian agriculture accounts for approximately 5 percent of the global CH_4 budget. Nelson et al. (2009) used field-level data collected by Pathak et al. (2005) with two global land-use data sets to assess the costs and benefits of a midseason drying. They found that, with one midseason drying, net revenue drops less than 5 percent, while GHG emissions drop by almost 75 million mt of CO_2e. The opportunity cost is US$1.20 per mt CO_2e, which is well below current carbon prices in European markets.

Case study: Mitigation within one irrigation system in the Philippines

Bohol Island, one of the largest rice-growing areas in the Visayas region of the Philippines, has experienced declining productivity and income from existing irrigation systems. The problem has been aggravated by the practice of unequal water distribution and unnecessary water use by farmers who insist on continuous flooding to irrigate their rice crop. The construction of a new dam was accompanied by a plan to implement a water-saving technology called alternate wetting and drying (AWD), developed by IRRI in cooperation with national research institutes. Visible success of AWD in pilot farms and specific training programs for farmers have helped to dispel the widespread misperception of possible yield losses in nonflooded rice fields. Adoption of AWD facilitated improved use of irrigation water and increased rice productivity. Using the methodology of the Intergovernmental Panel on Climate Change (IPCC), modification of water regime also can reduce methane emissions by almost 50 percent as compared to rice produced under continuous flooding. The Bohol case is an example of new technologies that increase the income of poor farmers while decreasing GHG emissions.

Suggested negotiating outcomes:

The two case studies demonstrate the potential for large reductions in rice production GHG emissions with relatively low opportunity costs and, in some cases, increases in productivity. Adapting the technologies to local conditions is necessary, and involving local farmers, extension agents, and research institutions in technology design and dissemination is critical. Measuring the reductions in GHG emissions can be done by using process methods supplemented with some field testing. Methane reduction from irrigated rice should be made eligible for offsets and other mitigation funding opportunities as an outcome of the Copenhagen negotiations.

Rice production also demonstrates the potential pitfalls of allocating Certified Emission Reductions (CERs) in the land-use sector. Water-saving techniques can reduce GHG emissions in a given area of rice land, but, in most cases, the saved water will then be used to irrigate more rice land or new crops in future seasons. Subsequently, emission savings are offset by emissions created in newly irrigated land. Ironically, if the saved water was channelled to other users, for example, in residential areas, one could rightfully claim CERs because of a net reduction in GWP caused by the mitigation project.

Increasing food production is an absolute necessity for the human population, and improved resource-use efficiencies are imperative to achieving this goal. As an agricultural research institution devoted to the increase in food production, IRRI proposes specific provisions for CER allocations in the land-use sector to converge the legitimate goals of food security and GHG mitigation in a Copenhagen agreement. Our suggestion is to compute for net GWP savings based on food production targets. As long as saved resources, namely water and fertilizers, are used to increase food production in a resource-efficient manner, it seems undue to account for new emissions as offsets or leakages of a mitigation project. ■

For Further Reading: G. C. Nelson et al., *India Greenhouse Gas Mitigation: Issues for Indian Agriculture,* IFPRI Discussion Paper (Washington, D.C.: International Food Policy Research Institute, forthcoming); G. Pan et al., "Storage and Sequestration Potential of Topsoil Organic Carbon in China's Paddy Soils," *Global Change Biology* 10 (2003): 79–92; H. Pathak et al., "Greenhouse Gas Emissions from Indian Rice Fields: Calibration and Upscaling Using the DNDC Model," *Biogeosciences* 2, no. 2 (2005): 113–123; X. Yan et al., "Global Estimations of the Inventory and Mitigation Potential of Methane Emissions from Rice Cultivation Conducted Using the 2006 IPCC Guidelines," *Global Biogeochemical Cycles,* forthcoming; X. Yan et al., "Statistical Analysis of the Major Variables Controlling Methane Emission from Rice Fields," *Global Change Biology* 11 (2005): 1131–1141.

Reiner Wassmann *(r.wassmann@cgiar.org) is the coordinator of the Rice and Climate Change Consortium at the International Rice Research Institute (IRRI) and senior scientist at Karlsruhe Research Center,* ***Yasukazu Hosen*** *(y.hosen@cgiar.org) is a senior researcher at the Japan International Research Center for Agricultural Sciences and a soil scientist at IRRI, and* ***Kay Sumfleth*** *(k.sumfleth@cgiar.org) is a Centrum für internationale Migration und Entwicklung (CIM) integrated expert and visiting research fellow at IRRI.*

International Food Policy Research Institute
2033 K Street, NW • Washington, DC 20006-1002 • USA
Phone: +1-202-862-5600 • Skype: ifprihomeoffice • Fax: +1-202-467-4439 • Email: ifpri@cgiar.org
IFPRI® **www.ifpri.org**

Agriculture and Climate Change: An Agenda for Negotiation in Copenhagen

Direct and Indirect Mitigation Through Tree and Soil Management

Brent M. Swallow and Meine van Noordwijk

Focus 16 • Brief 4 • May 2009

Many opportunities exist for mitigating greenhouse gas (GHG) emissions through better management of trees and soils. There is potential for both direct mitigation through better management of carbon in agricultural landscapes and indirect mitigation through reduced pressure on carbon stored in forests, peatlands, and wetlands. Effectively harnessing these opportunities will take bold action in climate change negotiations.

Mitigating GHGs through better farm management of soil and trees

The fourth assessment of the Intergovernmental Panel on Climate Change (IPCC) focuses on direct mitigation in agriculture, concluding that 90 percent of the technical potential for direct mitigation is through sequestration of soil carbon in developing regions, particularly in Southeast Asia, South America, East Asia, and Eastern Africa. The greatest opportunities for cost-effective mitigation are through changes in cropland and grazing land management, restoration of organic carbon to cultivated soils, restoration of degraded lands, and agroforestry.

In harnessing the potential for soil carbon sequestration, a major challenge has been the cost of measuring and monitoring soil carbon, with measurement costs possibly exceeding the market value of soil carbon enhancement. Fortunately, some promising solutions to this problem have been found. For example, researchers at the World Agroforestry Centre have developed techniques for estimating soil characteristics from the reflectance properties of soil samples. Combined with satellite imagery, these techniques can be used to generate soil carbon maps for large landscapes.

The IPCC report draws the somewhat surprising conclusion that agroforestry—the deliberate management of trees in agricultural landscapes—has less potential for cost-effective carbon storage on agricultural land than many other land-use practices. Other studies have reached different conclusions. An earlier IPCC report found large potential for carbon sequestration through improved management of existing agroforestry systems and through conversion of degraded lands into agroforestry. A 2005 review of the evidence from Africa found that improved fallow systems using agroforestry can sequester between 0.1 and 5.3 metric tons (mt) of carbon per hectare per year, while conservation farming systems without trees can only sequester 0 to 0.36 mt per hectare per year. One reason for this wide range of estimates is the large variety of farming systems that can be described as agroforestry.

An advantage of tree-based systems is that current technologies make it easier to estimate above-ground biomass than soil carbon. A study conducted on various land uses in the semiarid Sahel found a strong positive correlation between total biomass and soil carbon, with carbon in the soils 5 to 20 times higher than carbon in the trees. Correlations are similarly high in more humid areas, where a higher proportion of carbon is stored above ground. Thus the potential is high for estimating total soil and above-ground carbon from data on above-ground carbon, rainfall, and soil type. A further advantage of agroforestry systems is that they can generate substantial benefits through increased income and products (such as livestock feed, fuelwood, fruit, and medicines). Higher levels of soil carbon increase soil fertility and thus enhance agricultural production.

Suggested negotiating outcome: Provide financial and institutional support for a mix of early action, coordinated research, capacity building, and information sharing to enhance carbon storage in agricultural landscapes.

Given the wide range of systems that can be classified as agroforestry or conservation agriculture, more coordinated empirical studies are needed on biomass, soil organic carbon, and soil fertility implications of those systems in a range of circumstances. There is also a need to refine and expand the use of techniques for large-scale measurement of soil and biomass carbon.

Reducing pressure on forest resources

One of the greatest opportunities for agriculture to mitigate climate change is indirect—through reduced pressure on forest resources. The decision on Reduced Emissions from Deforestation in Developing Countries (REDD), adopted by the 13th Conference of Parties to the Framework Convention on Climate Change (FCCC), "... encourages Parties...to address the drivers of deforestation relevant to their national circumstances, with a view to reducing emissions from deforestation and forest degradation." In 2009, negotiations in the FCCC have focused on REDD-plus, which considers reduced emissions from deforestation and degradation, and enhancement of carbon stocks through sustainable forest management and afforestation.

Of the many drivers of deforestation, expansion of agriculture is most important. The Food and Agriculture Organization of the United Nations (FAO) estimates that in 2005, agricultural lands occupied almost 50 million square kilometers of the earth's surface, having increased by about 10 percent since the 1960s. Expansion of smallholder agriculture is a particularly important driver of deforestation in Sub-Saharan Africa, where food production per capita has stagnated despite agricultural area expanding by about 2 percent per year. The 2008 World Development Report showed that growth in agricultural production has relied primarily on expansion of farming area in Africa and on more intensive use of purchased inputs in Asia. Despite being heavily dependent on agriculture, most African countries invest low proportions of their national budgets on agricultural research and development.

If forests and woodlands are valued for the land they occupy, the timber that can be extracted, and the soil fertility they provide to extensive agriculture, then enhanced road access, more profitable land-use technologies, and stronger markets for food and fuel crops will increase pressure on forest resources. This appears to describe the major expansion of cattle ranching in the Amazon, cocoa production in the Guinea forests of West Africa, monoculture coffee growing in Vietnam, tobacco production in southern Africa, and oil palm production in Southeast Asia.

Different dynamics are possible, however. A review of evidence from around the developing world has concluded that technological progress in intensive agriculture and labor-intensive technological progress can, under specific circumstances of labor absorption, reduce pressure on forests. Equally important for reduced deforestation, however, are the development and enforcement of secure property rights and control of migration into forest margin areas. However, the potential negative effects of intensification on nitrous oxide and methane emissions needs to be evaluated, since small emissions of these more dangerous GHGs may offset part of the reductions in carbon dioxide emissions occurring through reduced deforestation. A doubly effective solution can be achieved when intensive production systems also sequester substantial amounts of carbon and have a tight nitrogen cycle, as is the case for multistrata agroforestry systems.

Suggested negotiating outcome: Strengthen the contribution of soil and tree management in agriculture for a more effective REDD or REDD-plus mechanism.

The agreement on REDD reached in FCCC negotiations in 2007 recognized the need to address the drivers of deforestation. Negotiators should go further to recognize the implications of agriculture as a dominant driver. A mechanism that encourages reduced emissions from all land uses would be the most effective means to address these interactions.

Trade-offs and opportunities for synergies

There are both trade-offs and opportunities for synergies between carbon stocks and private economic returns to land users. During the past 15 years, the Alternatives to Slash and Burn Programme (ASB) has examined the trade-offs between carbon stocks and private economic returns to land users in landscapes across the tropical forest margins. The most recent ASB information on trade-offs examines the opportunity costs of avoided deforestation at sites in Cameroon, Indonesia, Peru, and the Philippines. The results show positive but relatively low opportunity costs in terms of forgone income per ton of extra carbon. In other words, reduced emissions from deforestation can be cost effective but certainly not free.

Suggested negotiating outcome: Provide land users with real incentives to maintain carbon stocks.

While land-use changes that both increase carbon stocks and farmers' income are possible, farmers generally will have to accept trade-offs. In-kind or monetary payments should be provided directly to farmers who bear the costs of forgone development opportunities.

Threatened carbon pools

Peatlands and wetlands are important carbon pools that are under particular threat from agriculture. A controversial 2006 report showed massive GHG emissions from conversion of peat forests and from poor management of peat soils converted earlier. Subsequent studies have made some adjustments to those results and emphasized the uncertainty surrounding these estimates. In a 2007 report, Swallow and others show that conversion of peat forests in the Jambi province of Indonesia has generated large amounts of emissions at very low returns to farmers, often followed by land abandonment. Peatlands are found in many developing countries, with many other developing countries containing large areas of high-carbon wetlands facing similar threats from agricultural expansion.

Suggested negotiating outcome: Address emissions from past conversion of peat forests and poor management of peatlands in all countries with substantial peatland areas.

A mechanism that encourages reduced emissions from all land uses could accommodate the pressing need to reduce emissions from peatlands and wetlands.

IPCC Guidelines Provide an Accounting Base

The IPCC's guidelines for reporting emissions from agriculture, forestry, and other land uses are already being used by developed countries in their reports to the FCCC. Rather than develop a patchwork of rules for different aspects of land use, a comprehensive accounting system such as the IPCC guidelines should be applied in all countries. Otherwise issues such as leakage and additionality may be addressed through complex rules, resulting in high transaction costs and low effectiveness. This problem undermined the potential for afforestation projects under the Clean Development Mechanism.

Suggested negotiating outcome: Base accountability for the net emissions from all land use on existing IPCC agriculture, forestry, and other land-use guidelines. ■

For Further Reading: Food and Agriculture Organization of the United Nations (FAO), *Global Forest Resources Assessment 2005*, Forestry Paper No 147 (Rome: FAO, 2006); A. Hooijer et al., *PEAT-CO2, Assessment of CO2 emissions from drained peatlands in SE Asia*, Delft Hydraulics report Q3943 (Netherlands: Delft Hydraulics, 2006); Swallow, B. et al., *Opportunities for Avoided Deforestation with Sustainable Benefits*. An Interim Report by the ASB Partnership for the Tropical Forest Margins (Nairobi, Kenya: ASB Partnership for the Tropical Forest Margins, 2007); T. G. Vagen, R. Lal, and B. R. Singh, "Soil Carbon Sequestration in Sub-Saharan Africa: A Review," *Land Degradation and Development* 16: 53–71 (2005).

Brent M. Swallow *(b.swallow@cgiar.org) is a principal economist at the World Agroforestry Centre in Nairobi, Kenya and* ***Meine van Noordwijk*** *(m.vannoordwijk@cgiar.org) is a principal ecologist at the World Agroforestry Centre in Bogor, Indonesia.*

International Food Policy Research Institute
2033 K Street, NW • Washington, DC 20006-1002 • USA
Phone: +1-202-862-5600 • Skype: ifprihomeoffice • Fax: +1-202-467-4439 • Email: ifpri@cgiar.org
IFPRI® **www.ifpri.org**

Agriculture and Climate Change: An Agenda for Negotiation in Copenhagen

The Potential for Soil Carbon Sequestration

Rattan Lal

2020 Vision for Food, Agriculture, and the Environment

Focus 16 • Brief 5 • May 2009

Of the five principal global carbon pools, the ocean pool is the largest at 38.4 trillion metric tons (mt) in the surface layer, followed by the fossil fuels (4.13 trillion mt), soils (2.5 trillion mt to a depth of one meter), biotic (620 billion mt), and atmospheric pools (800 billion mt). If the fluxes among terrestrial pools are combined, annual total carbon flows across the pools average around 60 billion mt, with managed ecosystems (croplands, grazing lands, and plantations) accounting for 57 percent of that total. Thus, land managers have custody of more annual carbon flows than any other group.

What is carbon sequestration?

Carbon concentration in the atmosphere is increasing at the rate of about 4 billion mt (2 parts per million) per year, with transfer primarily from the fossil fuel, biotic, and soil pools. This increase is a double jeopardy. One, the loss of carbon from the terrestrial pools reduces the ecosystem services and goods that these systems provide. In particular, decline in soil quality adversely affects use efficiency of inputs, decreases agronomic yields, and exacerbates food insecurity. Two, increase in atmospheric pools accentuates global warming with the attendant impact on pole-ward shifts of ecosystems and the increase in frequency and intensity of extreme events including droughts, melting glaciers and Arctic ice sheet, rising sea level, and loss of biodiversity. One solution to this problem is to transfer atmospheric CO_2 into other long-lived pools (such as the soil and biotic pools); this is called carbon sequestration. Increasing carbon pools in the soil beyond a threshold level (about 1.2 percent in the surface layer) is essential to enhancing soil quality, increasing agronomic productivity, and improving quality of natural waters. The strategy of carbon sequestration in soils and biota is cost effective, safe, and has numerous co-benefits over leaving carbon in the atmosphere or sequestering it in geologic and oceanic strata. Biotic, or plant-based, sequestration is based on a natural process whereby CO_2 is photosynthesized into organic substances and stored for the long term in plant products and soil organic matter. The natural rate of photosynthesis in the global biosphere is about 120 billion mt of carbon per year. Fossil fuel combustion emits about 8 billion mt of carbon annually, and deforestation and land-use conversion emit another 1.6 billion to 2 billion mt of carbon per year, for a total of 9.6 to 10.8 billion mt of carbon emissions per year. Thus, if roughly 8 percent of the carbon being photosynthesized by the biosphere is retained within the soil and biotic pools, the global carbon budget would be balanced.

The technical potential for soil carbon sequestration

Soil organic carbon has been depleted through (1) the long-term use of extractive farming practices and (2) the conversion of natural ecosystems (such as forest lands, prairie lands, and steppes) into croplands and grazing lands. Such a conversion depletes the soil organic carbon pool by increasing the rate of conversion of soil organic matter to CO_2, thereby reducing the input of biomass carbon and accentuating losses by erosion. Most agricultural soils have lost 30 to 40 mt of carbon per hectare, and their current reserves of soil organic carbon are much lower than their potential capacity.

Soil carbon sequestration involves adding the maximum amount of carbon possible to the soil. The technical potential for this process is higher in degraded/desertified soils and soils that have been managed with extractive farming practices than it is in good-quality soils managed according to recommended management practices (RMPs). Thus, converting degraded/desertified soils into restorative land and adopting RMPs can increase the soil carbon pool. While no single technology is appropriate for all soils, climates, or cropping and farming systems, the goal is to identify site-specific technologies that create a positive soil carbon budget. The rate of soil carbon sequestration through the adoption of RMPs on degraded soils ranges from 100 kilograms per hectare (kg/ha) per year in warm and dry regions to 1,500 kg/ha per year in cool and temperate regions. A recent estimate of the technical potential of soil organic carbon sequestration through adoption of RMPs for world cropland soils (1.5 billion hectares) is 0.4 billion to 1.2 billion mt of carbon per year.

Examples of soil and crop management technologies that increase soil carbon sequestration include

- no-till (NT) farming with residue mulch and cover cropping;
- integrated nutrient management (INM), which balances nutrient application with judicious use of organic manures and inorganic fertilizers;
- various crop rotations (including agroforestry);
- use of soil amendments (such as zeolites, biochar, or compost); and
- improved pastures with recommended stocking rates and controlled fire as a rejuvenate method.

Another good strategy for soil carbon sequestration is the restoration of degraded/desertified soils (about 2 billion hectares), which can be achieved through afforestation and reforestation. The technical potential of soil carbon sequestration through restoration of degraded/desertified soils is 0.6 billion to 1 billion mt of carbon per year. The establishment of energy plantations can also improve ecosystem carbon pools. It is estimated that afforestation and establishment of energy plantations can offset 25 billion mt of carbon between 2000 and 2050.

The technical potential of carbon sequestration in world soils may be 2 billion to 3 billion mt per year for the next 50 years. Thus, the potential of carbon sequestration in soils and vegetation together is equivalent to a draw-down of about 50 parts per million of atmospheric CO_2 by 2100.

Soil carbon in peatlands

One particularly important and unique soil carbon pool is in peatlands. Peat is formed when plant material in marshy areas is kept from decaying by acidic conditions. Draining and burning peatlands is a significant source of CO_2 emissions. Restoration of wetlands and avoiding cultivation of peatland can convert these soils from a large source to a vast carbon sink. Drained and cultivated peat lands decompose and subside at the rate of 1 to 2 centimeters per year. The rate of soil carbon sequestration in restored peatlands may be greater than 1 mt of carbon per hectare per year. This rate of sequestration is above the emission avoidance rate through inundation of wetlands because of the decomposition of cultivated peat. About 400 million hectares of peatlands in the world can sequester 0.4 billion mt of carbon per year.

The economic potential for soil carbon sequestration

One way to think of soil carbon is as a commodity. It can be produced and, if carbon markets exist, traded like any other farm produce. Additional income can be an important incentive for the resource-poor farmers in developing countries to invest in soil restoration and adopt RMPs. The economic potential may be as much as 60 percent of the technical potential, or 1.2 to 2.0 billion mt of carbon per year. Furthermore, measuring and monitoring protocols of change in carbon pools at the landscape, farm, and regional scales are available to facilitate carbon trading.

The greatest potential for sequestration is in the soils of those regions that have lost the most soil carbon. These are the regions where soils are severely degraded and have been used with extractive farming practices for a long time. Among developing countries, these regions include Sub-Saharan Africa, South and Central Asia, the Caribbean, Central America, and the Andean regions. Most soils have a technical or maximum sink capacity of 20 to 50 mt of carbon per hectare that can be sequestered over a 20-to-50-year period.

Soil carbon sequestration can enhance productivity and resilience

Increases in soil organic material have important productivity and resilience benefits. These benefits include improvement in soil quality, increase in use efficiency of inputs, reduction in soil erosion and sedimentation, decrease in nonpoint source pollution, and lower rates of anoxia or hypoxia (dead water) in coastal ecosystems. Global food security cannot be achieved without restoring the quality of degraded soils, for which soil carbon sequestration is an essential prerequisite.

Soil carbon sequestration is a win-win strategy. It mitigates climate change by offsetting anthropogenic emissions; improves the environment, especially the quality of natural waters; enhances soil quality; improves agronomic productivity; and advances food security. It is a low-hanging fruit and a bridge to the future, until carbon-neutral fuel sources and low-carbon economy take effect.

Suggested negotiating outcomes:

Carbon sequestration in soils and plants is the only strategy that can remove carbon from the atmosphere and, over time, reduce atmospheric concentration of CO_2. Initiatives to support reduced emissions from deforestation (REDD) are well underway. Funds for soil carbon mitigation should also be made available. Support should be provided for

- crop mixes to include more plants that are perennial or have deep-root systems in order to increase the amount of carbon stored in the soil;
- cultivation systems that leave residues and reduce tillage, especially deep tillage, in order to encourage the buildup of soil carbon;
- shifting land use from annual crops to perennial crops, pasture, and agroforestry in order to increase both above- and below-ground carbon stocks; and
- activities that restore degraded and desertified soils and ecosystems, especially those affected by accelerated erosion, salinization, and nutrient depletion.

Carbon offset payments should be allowed for carbon sequestered in soils where low-cost monitoring is available. Funds for the development of these monitoring systems should be part of any outcome.

Paying resource-poor farmers and smallholders in developing countries for soil carbon sequestration would contribute to GHG mitigation, provide much needed resources to support development and adaption of improved crop technologies, and reduce rural poverty. ■

For Further Reading: R. Lal. "Challenges and Opportunities in Soil Organic Matter Research," *European Journal of Soil Science* 60 (2009): 158–160; R. Lal, "Enhancing Crop Yield in Developing Countries through Restoration of Soil Organic Carbon Pool in Agricultural Lands," *Land Degradation and Development* 17 (2006): 197–209; R. Lal, "Soil Carbon Sequestration Impacts on Global Climate Change and Food Security," *Science* 304 (2004): 1623–1627.

Rattan Lal *(lal.1@osu.edu) is Director of the Ohio State University's Carbon Management and Sequestration Center and Professor of Soil Science in the School of Environment and Natural Resources.*

International Food Policy Research Institute
2033 K Street, NW • Washington, DC 20006-1002 • USA
Phone: +1-202-862-5600 • Skype: ifprihomeoffice • Fax: +1-202-467-4439 • Email: ifpri@cgiar.org
IFPRI® **www.ifpri.org**

Mitigating Greenhouse Gas Emissions from Livestock Systems

M. HERRERO AND P. K. THORNTON

2020 VISION FOR FOOD, AGRICULTURE, AND THE ENVIRONMENT

FOCUS 16 • BRIEF 6 • MAY 2009

Livestock—poultry, small ruminants (such as goats and sheep), cattle, and pigs—provide many benefits for human well-being. Livestock production systems, especially in developing countries, are changing rapidly in response to population growth, urbanization, and growing demand for meat and milk. The need for action by all sectors to mitigate climate change adds additional complexity to the already considerable development challenges these systems face.

Some livestock production systems use large quantities of natural resources and also produce significant amounts of greenhouse gas emissions (GHGs). Since the demand for meat and milk is increasing, the question is whether cost-effective mitigation options exist to meet them within equitably negotiated and sustainable GHG emission targets. In fact, emissions from livestock systems can be reduced significantly through technologies, policies, and the provision of adequate incentives for their implementation. The objective of this policy brief is to highlight options to mitigate GHGs from livestock industries and to suggest key negotiating outcomes for including livestock in the Copenhagen meetings.

The global livestock industry

Livestock systems occupy 45 percent of the global surface area with a value of at least $1.4 trillion. Livestock industries are a significant source of livelihoods globally. They are organized in both short and long market chains that employ at least 1.3 billion people globally and directly support the livelihoods of 800 million poor smallholder farmers in the developing world.

Livestock are an important source of nourishment. Livestock products contribute 17 percent of calorie consumption and 33 percent of protein consumption globally. The level of consumption of milk and meat in the developed world is at least five times higher than in the developing world. However, in developing countries the demand for livestock products is rising rapidly, mainly as a consequence of increased human population and rapidly increasing incomes, primarily in Asia. Growth in milk and beef production is also becoming important in parts of Africa. It is projected that growth in poultry and pig production will be adequate to satisfy the demand.

For the poor, increased consumption of livestock products has positive effects on mortality and the cognitive development of children. At the same time, the sale of livestock products can increase smallholders' incomes.

Keeping livestock can be an important risk-reduction strategy for vulnerable communities. And livestock are important providers of nutrients and farm traction in smallholder systems.

Of the planet's 1.3 billion poor who live on less than a dollar a day, at least 90 percent are located in Asia and Sub-Saharan Africa, and 60 percent of those depend on livestock for some part of their livelihood. Climate change is likely to have major effects on poor livestock keepers and on the ecosystems on which they depend. These impacts will include changes in the productivity of rainfed crops and forage; reduced availability of water and widespread water shortages; and changes in the severity and distribution of important human, livestock, and crop diseases. Major changes can thus be anticipated in livestock systems, including the mix of species raised, crops grown, and feed resources and feeding strategies.

Livestock and GHG emissions

Livestock contribute 18 percent of global anthropogenic GHG emissions (FAO 2006). The main sources and types of GHGs from livestock systems are methane production from animals (25 percent), carbon dioxide (CO_2) from land use and its changes (32 percent), and nitrous oxide (N_2O) from manure and slurry management (31 percent).

The systems for producing different kinds of livestock are highly diverse, which results in large differences in the associated GHG emissions per kilogram produced in different regions. The impacts of livestock production on GHG emissions have been widely discussed, particularly those associated with rapidly expanding industrial livestock operations in Asia and those linked to deforestation in Latin America. Nevertheless, in smallholder crop-livestock, agropastoral, and pastoral-livestock systems, livestock are one of a limited number of broad-based options to increase incomes and sustain the livelihoods of people who have a limited environmental footprint. By diversifying risk and increasing assets, livestock increase the resilience of vulnerable poor people, who are subject to climatic, market, and disease shocks. Given that almost all human activity is associated with GHG emissions, those from livestock in these systems are relatively modest, compared with the contribution that livestock make to the livelihoods of a huge number of people. This complex balancing act of resource use, GHG emissions, and livelihoods must be clearly understood and taken into account when designing mitigation strategies to offset the effects of livestock on the environment. Farmers should be provided incentives or offset payments for adopting livestock systems that reduce emissions yet maintain their livelihoods.

GHGs emitted by livestock systems can be significantly reduced

GHG emissions in livestock systems can be reduced through technologies, policies, and incentives. The important ways are managing the demand for livestock products, intensifying the diets of ruminants, using more productive livestock breeds, or shifting species.

Consumption of livestock products per capita has increased over the last few decades in the developed world, and recent evidence suggests that this level of consumption in some countries increases the risk of health problems. In these countries demand is met by local production in intensive systems or by direct imports of livestock products. In both cases, this demand affects land-use practices and use of resources in the developing world that are associated with significant GHG emissions. Reducing demand for livestock products in the developed world could lead to healthier people and also reduce pressures on land and natural resources in developing countries. This could lead to significant reductions in CO_2 and methane emissions.

The amount of methane produced per unit of animal product can be reduced by feeding better quality diets to ruminants. This increased efficiency could be achieved through improved land-use management with practices such as improved fodder technologies (development of fodder banks, improved pasture species, use of legumes, and others) and supplementation with crop by-products. These practices, which are cost effective and available in developing countries, can increase milk production, improve the efficiency of methane production, and, together with reductions in the number of animals, help mitigate methane emissions from ruminant systems. Other options include manipulation of rumen microflora and use of feed additives, as practiced in some parts of the developed world, although reductions are only likely to be on the order of 10 percent at best. In the developing world, many low-producing animals could be replaced with fewer but better-fed animals, thus reducing total emissions while maintaining or increasing the supply of livestock products. This will require changing breeds or implementing cross-breeding schemes. Switching livestock species to better suit particular environments is a strategy that could yield higher productivity per animal for the resources available. Also, switching from cows, sheep, and goats to pigs and poultry could lead to reduced methane emissions, although it could also increase the demand for grains. More research is required to understand the effects of these trade-offs between species.

Regulations are required to reduce N_2O emissions from manures. They are of particular importance for managing excreta in the developing world and for slurry and manure applications from cattle in the developed world. In the developing world, regulatory frameworks for manure management in poultry and pig industrial units are necessary to reduce emissions.

Grazing systems can enhance the removal of CO_2 from the environment

Carbon can be sequestered (or, captured) from the atmosphere via improved management. Any practice that increases the photosynthetic uptake of carbon or slows the return of stored carbon to CO_2 via respiration, fire, or erosion will increase carbon reserves, thereby sequestering carbon. Significant amounts of soil carbon could be stored in rangelands or in silvopastoral systems through practices suited to local conditions. This would not only improve carbon sequestration but could also turn into an important diversification option for sustaining livelihoods of smallholders and pastoralists through collection of payments for ecosystem services.

Finally, livestock is integrally linked to crop production in the developing world. Crops and residues from agricultural lands are used to feed livestock, and manure is a crucial source of nutrients for crop growth and as fuel in crop–livestock systems. Crop residues can also be used as a source of fuel, either directly or after conversion to fuels such as ethanol or diesel. While these bioenergy feedstocks still release CO_2 upon combustion, the carbon is of recent atmospheric origin (via photosynthesis), rather than from fossil carbon. The net benefit of these bioenergy sources to the atmosphere is equal to the fossil-derived emissions displaced, less any emissions from producing, transporting, and processing. CO_2 emissions can also be avoided through agricultural management practices that forestall the cultivation of new lands now under forest, grassland, or other nonagricultural vegetation.

Suggested negotiating outcomes:

- Fund the implementation of effective strategies to mitigate the impacts of livestock in the developing world, while balancing the need to produce food (in the form of livestock products) in non-arable areas and to enable the vulnerable poor to continue to earn a living from livestock keeping.
- Fund mechanisms for developing countries to improve ruminant feeding and research for better understanding of the trade-offs between improved feeding practices and reduced animal numbers in different parts of the world.
- Fund research to elucidate the effects of changing breeds or species on the supply of animal products, smallholder incomes, and the best use of land for food production, while still meeting carbon targets and not compromising smallholder livelihoods.
- Fund the implementation of mitigation techniques from manure management in industrial pig and poultry systems in developing countries to reduce N_2O emissions from livestock systems.
- Fund the implementation of schemes to collect payments for agro-ecosystem services in selected rangeland systems to increase their contribution as a carbon sink and to provide income diversification options for pastoralists in developing countries. ■

For Further Reading: M. Herrero et al. "Drivers of change in crop–livestock systems and their potential impacts on agro-ecosystems services and human well-being to 2030" (Nairobi: International Livestock Research Institute, 2009); P. K. Thornton et al. "Mapping climate vulnerability and poverty in Africa," http://www.dfid.gov.uk/research/climate-change.asp (2006); H. Steinfeld et al. *Livestock's long shadow: Environmental issues and options* (Rome: Food and Agriculture Organization of the United Nations, 2006).

***M. Herrero** (m.herrero@cgiar.org) and **P. K. Thornton** (p.thornton@cgiar.org) are systems scientists at the International Livestock Research Institute, Nairobi, Kenya.*

International Food Policy Research Institute
2033 K Street, NW • Washington, DC 20006-1002 • USA
Phone: +1-202-862-5600 • Skype: ifprihomeoffice • Fax: +1-202-467-4439 • Email: ifpri@cgiar.org
IFPRI® **www.ifpri.org**

The Role of Nutrient Management in Mitigation

Helen C. Flynn

Focus 16 • Brief 7 • May 2009

Nitrous oxide (N_2O) emissions from soils are responsible for about 3 percent of greenhouse gas (GHG) emissions, which cause climate change, and contribute approximately one-third of non-CO_2 agricultural GHG emissions. N_2O is produced by microbial transformations of nitrogen in the soil, under both aerobic and anaerobic conditions. Therefore, emissions are often directly related to nutrients added to the soil in the form of mineral fertilizers and animal manure. These additions can be vital in maintaining soil fertility and crop production; about half of the world's population is dependent on food produced strictly because of mineral fertilizer inputs. However, the additions are also highly inefficient, leading to nitrogen losses via leaching, volatilization, and emissions to the atmosphere. By helping to maximize crop-nitrogen uptake, improved nutrient management has a significant and cost-effective role to play in mitigating GHG emissions from agriculture. Nutrient management can also help reduce methane (CH_4) emissions from rice production and increase carbon sequestration in agricultural soils.

Mitigation strategies

Mineral fertilizer use

Management strategies to improve the nitrogen use efficiency of crops, thereby reducing fertilizer requirements and associated GHG emissions, focus on fertilizer best management practices. These practices are based on the principle of the "right source, at the right rate, at the right time, and with the right placement." Such practices concentrate on the following:

Application type

- Researchers have argued that urea-based fertilizers lead to higher N_2O emissions than ammonia or nitrates do, but the most recent reviews suggest that both environmental factors, such as soil conditions and climate, and management factors, such as tillage, also play key roles in determining the proportion of applied nitrogen lost as N_2O. These confounding variables prevent valid comparisons between fertilizer types. Also, some forms of nitrogen fertilizer may reduce N_2O emissions but not improve overall nitrogen use efficiency due to other nitrogen losses, for example, through leaching. Although research into nitrogen sources has given mixed results, it is clear that "balanced" fertilization (that is, balancing nitrogen applications with other required nutrients, including phosphorus, potassium, and sulphur) is a major way of improving nitrogen use efficiency.

Application rates

- Appropriate nitrogen application rates are required to limit the build-up of nitrates in soil, which can accumulate when more nitrogen is applied than the crop demands at that time. Cutting nitrogen application rates below economic optimums risks long-term decline in soil productivity, a problem already occurring in places like Africa and parts of India that are chronically under-fertilized.

  Managing nitrogen fertilizer levels is challenging because appropriate application rates will differ for each agroecosystem and growing season. The exact relationship between nitrogen input and N_2O emissions is debatable, but many studies have suggested that when an agronomic nitrogen-threshold level—an amount based on the ecosystem uptake capacity determined by field measurements—is exceeded, N_2O emissions increase dramatically.

Application timing

- Nitrogen applications that are carefully timed to maximize crop uptake reduce application rates and N_2O emissions without decreasing crop yield. Applications should be avoided prior to planting and, instead, concentrated in the initial crop development phase at the time of, and shortly after, planting in order to maximize crop uptake and minimize nitrogen loss from the system.

Application placement

- Greater proportions of applied nitrogen are generally lost if fertilizer is applied at the surface, although this may be in the form of ammonia-volatilization rather than N_2O emissions. Researchers disagree as to the effect of application depth, which appears to be strongly influenced by the tillage regime, among other factors.

Best practice for fertilizer use is dependent, to a certain extent, on the exact agroecosystem under consideration, its management regime, and environmental factors. Thus, management plans need to reflect local conditions. Precision farming systems are already available to ensure farmers can draw up careful plans, and the most advanced systems can reduce fertilizer usage by about one-third. For small-scale farmers in the developing world, who have no access to modern farming equipment, the best solution for improving fertilizer practices is to increase access to independent advice from local experts such as research institutes.

Another possible mitigation strategy is the wider use of fertilizer additions such as controlled-release coatings and nitrification inhibitors, which also reduce CH_4 emissions from fertilized rice paddies. Controlled-release or enhanced-efficiency fertilizers generally work by controlling the speed at which fertilizer, or a coating applied to it, dissolves in soil water. By affecting the timing of nitrogen release from fertilizer, these compounds have the potential to reduce the loss of nitrogen and therefore improve nitrogen use efficiency.

Similarly, soluble fertilizers formulated with inhibitors reduce or block the conversion of nitrogen species by affecting specific types of microbes involved. This helps to keep nitrogen in the form of ammonium longer, encouraging uptake by crops and helping to prevent N_2O emissions from either nitrification or denitrification. Some inhibitors can be more than 90 percent effective in reducing N_2O emissions.

Mineral fertilizer production, distribution, storage, and application currently contribute approximately 2 percent of total global GHG emissions. Further research could clarify the best fertilizer additions to include under specific circumstances and develop new nitrogen-related products, including, for example, smart delivery mechanisms that are driven by factors related to temperature, water, or biotic properties, such as the host plant. Better understanding of the relationship between N_2O emissions and extreme weather events may also become increasingly important as our climate changes.

Organic nitrogen sources

With synthetic fertilizers inducing N_2O emissions from soils, requiring energy, and producing GHG emissions during their manufacture, another key mitigation strategy is to make better use of existing organic sources of nutrients, including animal manure, crop residues, and nitrogen-fixing crops such as legumes. Such organic nitrogen sources may also contribute to increasing carbon sequestration in soils.

Animal manure

- As well as reducing mineral fertilizer requirements and the GHG emissions associated with their manufacture, using animal waste may also reduce soil N_2O emissions. While nitrogen losses from liquid slurry are generally higher than from mineral fertilizers, many researchers argue that solid manures reduce emissions. Research also suggests that organic manures do not cause the spikes in emissions that occur with mineral fertilizers if there is heavy rainfall around the time of application, meaning manures can significantly mitigate N_2O emissions during wet growing seasons.

Crop residues

- Incorporating post-harvest plant remains into soil can increase levels of soil organic matter thereby assisting in soil carbon storage. However, these benefits can be offset by increased N_2O emissions in some cases, so residues with high nitrogen content could be composted prior to incorporation to minimize negative effects. Impacts are also dependent on tillage management and other fertilizer additions.

Nitrogen-fixing crops

- Introducing crops such as clover and other legumes into rotations can reduce fertilizer requirements by adding biologically fixed nitrogen into soils. While this tends to raise background emissions of N_2O from the soil, it can mitigate total emissions over a longer term (accounting for reductions in N_2O emissions from mineral fertilizers applied to the other crops in rotation). Planting nitrogen-fixing trees during fallow periods is another good option.

Much of the research into the effects of organic additions to date has focused either on N_2O emissions or on carbon sequestration, making it difficult to ascertain the net impact in terms of global warming potential (GWP). Further research in this area would be valuable, especially since organic additions are among the cheapest mitigation strategies available. The integrated use of organic and synthetic fertilizers may be the best option for improved soil fertility and crop production, as organic sources make it harder to synchronize nitrogen release with crop demands and may increase N_2O emissions in comparison with synthetic fertilizers under some circumstances. Particularly in many developing countries where there is pressure on organic resources—for example, in places where farmers need to use animal dung as fuel—agroforestry should be promoted. Planting trees will offer farmers an alternative fuel supply and benefit soils and crops.

Mitigation potential

More efficient use of mineral fertilizers is highly achievable; countries such as China and India—the two largest consumers of synthetic nitrogen—currently have much lower crop-use efficiencies than areas like Europe, where fertilizer use has declined in recent years. Policies regarding heavy subsidies for nitrogen fertilizers have contributed to this inefficiency, and implementing fertilizer best management practices can help reduce inefficiency while also (1) reducing GHG emissions and other environmental damage, such as nitrification of waterways, (2) improving nutrient balance and efficiency, (3) lowering fertilizer costs to farmers, and (4) freeing up government spending for more beneficial projects. Reductions in the use of mineral fertilizers also have additional benefits in terms of GHG savings from their manufacture and transportation, which can be vastly higher from inefficient coal-powered plants in countries such as China and Russia, compared with the most advanced plants. Upgrading all fertilizer production plants to the best modern standards can also significantly reduce energy requirements and N_2O emissions.

Currently, crop production per unit of nitrogen applied is falling in many countries, such as China, because crops are vastly overfertilized, while soils in parts of Africa and India still suffer from chronic nutrient deficiency. This imbalance needs to be addressed. Increased and better use of organic material for fertilization can assist with this while also mitigating indirect emissions from synthetic fertilizer production, improving soil quality through more balanced nutrition, and sequestering more carbon into soil systems.

Suggested negotiating outcomes:

Sufficiently robust practices that should be implemented and rewarded with offset payments under current conditions include fertilizer best management practices (where they have not already been implemented); the use of controlled-release fertilizers and nitrification inhibitors; applying animal manure to and incorporating crop residues into the soil in areas where soil organic matter is declining; and introducing nitrogen-fixing crops into intensive rotations. Although not strictly a nutrient-management issue, agroforestry would also be significantly beneficial.

Further research funding could be targeted at better understanding the net effects of introducing nitrogen-fixing crops into less heavily fertilized crop rotations, using organic amendments in areas where soil organic matter is already well maintained, and finding new nitrogen products along the lines of controlled-release fertilizers. Better long-term field trials need to be established so that soil carbon gains and reduced nitrogen emissions can be quantified for the purpose of offset payments. ■

For Further Reading: J. Bellarby et al., *Cool Farming: Climate Impacts of Agriculture and Mitigation Potential* (Amsterdam: Greenpeace International, 2008); R. L. Desjardins et al., "Management Strategies to Sequester Carbon in Agricultural Soils and to Mitigate Greenhouse Gas Emissions," *Climatic Change* 70 (2005): 283–297; C. S. Snyder et al., *Greenhouse Gas Emissions from Cropping Systems and the Influence of Fertilizer Management* (Norcross, Ga., U.S.A: International Plant Nutrition Institute, 2007).

***Helen C. Flynn** (h.c.flynn@abdn.ac.uk) is a research fellow at the Institute of Biological and Environmental Sciences, School of Biological Sciences, University of Aberdeen, Scotland, UK.*

International Food Policy Research Institute
2033 K Street, NW • Washington, DC 20006-1002 • USA
Phone: +1-202-862-5600 • Skype: ifprihomeoffice • Fax: +1-202-467-4439 • Email: ifpri@cgiar.org
IFPRI® **www.ifpri.org**

Monitoring, Reporting, and Verification Methodologies for Agriculture, Forestry, and Other Land Use

SEAN SMUKLER AND CHERYL PALM

FOCUS 16 • BRIEF 8 • MAY 2009

Facilitating carbon sequestration in terrestrial ecosystems could provide a significant amount of atmospheric carbon dioxide (CO_2) abatement, which is necessary to limit global temperature increases to only 2 degrees Celsius in the next century until more permanent mitigation strategies are instituted. With relatively small investments, greenhouse gas (GHG) emissions could be offset dramatically by management practices such as planting trees, reducing deforestation, midseason draining of irrigated rice, improving nitrogen fertilization efficiency, and increasing organic matter inputs to agricultural soils. Together these types of practices could add up to more than 25 percent of the combined near-term abatement strategies (including energy efficiency and low-carbon energy supply) required to stabilize emissions.

While most terrestrial management potential is based on reduced deforestation and degradation (REDD), no one program can be effective in isolation. It is crucial to recognize that there are multiple competing uses for land and that maximizing GHG mitigation is not likely to be achieved with carbon-based financial incentives alone, particularly if incentives do not reach those most responsible for land management. Nearly 90 percent of the potential for terrestrial carbon capture can be found in the developing world, where land managers are largely poor farmers on small plots of land. It is imperative that these farmers be involved in carbon mitigation strategies, but dealing with numerous smallholders is an enormous challenge because planning, monitoring, reporting, and verifying mitigation creates transaction costs for carbon contracts that can be prohibitively expensive. It is therefore critical for the international community to immediately invest in the research and development of innovative methodologies to reduce transaction costs by increasing the effectiveness of monitoring, reporting, and verification for Agriculture, Forestry and Other Land Use (AFOLU) projects, particularly for smallholder agriculture in tropical regions.

AFOLU projects are a win–win for mitigation and adaptation

There is enormous potential for AFOLU projects to create carbon mitigation programs that can help increase the adaptive capacity of those most at risk to climate change, while also improving livelihoods, preserving biodiversity, and ensuring the sustainability of both the farming system and terrestrial mitigation projects. Management practices that increase organic inputs, such as cover crops, stover management, and reduced tillage can sequester on average between 0.5-1 metric tons (mt) of carbon per hectare (ha) per year in the soil for up to 20 years and also increase soil water-holding capacity and soil fertility. The integration of trees into agricultural systems through agroforestry practices could also sequester up to 5 mt of carbon per ha per year for 20 to 30 years while at the same time generating alternative sources of income, providing organic nutrients, preventing erosion, and producing animal fodder.

More efficient use of nitrogen on the farm could reduce nitrous oxide emissions, which account for 2 percent of the total agricultural technical mitigation potential. More importantly, it could also increase the benefit-to-cost ratio of production and reduce the need for further agricultural expansion. Partnering with local communities so that they receive tangible benefits from AFOLU projects will help prevent leakage and ensure the long-term success of mitigation. To enable the participation of smallholders, transaction costs must be reduced through innovative, cost-effective monitoring, reporting, and verification methods.

Protocol development for AFOLU projects

Developing a consistent set of streamlined, cost-effective, and reliable protocols for project development and monitoring, reporting, and verifying them is critical to the success of AFOLU projects. There has been significant progress made in methodologies that capitalize on a combination of remote sensing techniques and ground-based inventories and that could be adopted by the United Nations Framework Convention on Climate Change (UNFCCC). The challenge is to improve these methods even further, namely by integrating existing protocols with newer cost-effective techniques as simply as possible.

Improving remote sensing capabilities

Remote sensing through aircraft- or satellite-based data acquisition will inevitably be an important component for monitoring AFOLU projects, but to what extent is still uncertain. With increased funding for research and development, remote sensing could substantially change the costs and reliability of monitoring mitigation projects and enable greater participation even from small-scale agriculture. The primary tool to map and estimate land cover or land use at the regional level could be a low-cost, readily available coarse-to-medium-resolution satellite imagery (250m–1km to 10m–60m of resolution). While there are established methods for using historical Landsat satellite images (30 m x 30 m spatial resolution)—which are now freely available, as the primary observation source for developing regional scale baselines by monitoring land-use change—they are not yet sufficient for farm-scale monitoring. Fine-resolution platforms (less than 5 meters), such as IKONOS and QuickBird, that create digital data of large-scale airphoto quality could provide the detail required for farm-scale monitoring; these images are too costly for wide-scale monitoring but could be used as training sets or verification of land cover. Clear guidance must be provided and methods must be defined for integrating, when possible, satellite observations from other optical systems (for example, RapidEye, SPOT, ASTER, or CBERS) or microwave satellite sensors (including JERS or ERSSAR) that could increase the resolution of land-use analyses by overcoming the limitations of optical sensors, such as darkness or cloud cover.

Investment in the research and development of new technologies and ways to rapidly incorporate their capabilities into protocols is also necessary. Newer technologies such as radar, Light Detection and Ranging (LiDAR), hyperspectral imaging, and combinations of these data show great promise for remotely estimating vegetation biomass. However, they are currently too expensive to be employed on a wide scale. Making these new technologies operational for carbon contracts in agricultural communities would require substantial reductions in acquisition costs. Over the next decade, there will be a number of sensors that could dramatically improve the capacity to observe changes in forest cover and farm management, including platforms from Brazil that will provide free data to developing nations (for example, CBERS-5 and CBERS-6), several types of commercial sensors, and spaceborne LiDAR.

Improving ground-based inventory techniques and capacity

While remote sensing has the potential to provide accurate regional-level monitoring of mitigation projects through land-cover and land-use change, it will require on-the-ground inventories to calibrate and verify analyses (as will any monitoring of farm-level projects). Reducing the cost of these inventories will require capital investment in two main areas: data collection and analysis and capacity building. Investments are needed to improve (1) the availability of regional data for estimating carbon stored in trees, (2) the development of new soil-analysis technology that will estimate soil carbon stocks to some specified depth, (3) the implementation of innovative data management technologies at farm and community levels, and (4) the connection of such information to carbon and nitrogen models for both the local and regional levels. Data are being collected to digitally map soils worldwide, beginning with Sub-Saharan Africa (through the African Soil Information System), and will substantially improve the potential to model carbon dynamics and crop-nitrogen requirements across the landscape. The digital soil mapping project is largely made possible by new soil-analytical techniques that use visible-near-infrared reflectance (Vis-NIR) spectroscopy, which allows for rapid, inexpensive evaluation of soil carbon.

Other new soil-analysis technologies that have the potential to revolutionize carbon assessments include laser-induced breakdown spectroscopy (LIBS) and inelastic neutron scattering (INS). In addition to being field deployable, INS has the added advantage of sampling in situ and does not require labor-intensive bulk density sampling. Investing in the development and purchase of field-deployable soil-analysis equipment could significantly increase monitoring capabilities and reduce costs. Improved field sampling methods could be further enhanced by investment in the development of low-cost, handheld global positioning systems. These systems should have the capability to input inventory data and send them via text message to centralized data management centers, a process that is being piloted in remote medical clinics in a number of developing countries. These innovative tools will help define carbon models for local conditions so they can estimate carbon- and nitrogen-dynamics for AFOLU projects over the duration of mitigation commitments.

A large national and subnational carbon-project workforce must be developed to meet the demands for high-quality, cost-effective monitoring, reporting, and verification. Within the countries where AFOLU projects are to be implemented, training nationals in each sector of carbon-based projects could significantly reduce project costs over time and create a workforce capable of effectively managing natural resources over the long term. Building the capacity of national agricultural extension could combine monitoring capabilities with better nitrogen management and increased crop production. Professional school, colleges, and universities need resources to implement the training required for such a workforce.

Investing in national and subnational accounting systems will create the transparency, credibility, and efficiency required to attract private carbon investment. If done well, this system could provide the institutional infrastructure to document national-level forest inventories while at the same time accounting for community-level agricultural, agroforestry, or afforestation/reforestation projects. Integrated systems for accounting and registry could substantially reduce the transaction costs of projects dealing with smallholders and enable greater overall participation in mitigation projects.

Suggested negotiating outcomes:

Terrestrial carbon mitigation could play a substantial role in an overall strategy to avoid dangerous levels of climate change over the next century. To maximize this mitigation potential, AFOLU projects require streamlined, cost-effective protocols for monitoring, reporting, and verification that facilitate large-scale participation. The technology to link remotely sensed soil and vegetation analysis to ground-based sampling exists, but pilot studies are necessary to demonstrate the capability, reliability, and affordability of such integrated analysis to the carbon community. Regional data collection, analysis, and modeling for biomass and soil carbon should be funded, as should research and development of protocols that maximize current remote sensing capabilities and new, low-cost, remote-sensing platforms. Once accepted, widespread investment in technology transfer, national and subnational accounting systems, and training for developing countries may be the most cost-effective means of meeting global near-term abatement goals. ■

For Further Reading: F. Achard et al., *Use of Satellite Remote Sensing in LULUCF Sector*, Background Paper, IPCC Expert Meeting, May 13-15 (Jena, Germany: Land Cover Project Office, 2008); P. Smith, et al., "Agriculture," in B. Metz, et al., eds., *Climate Change 2007: Mitigation, Contribution of Working Group III to the Fourth Assessment Report of the Intergovernmental Panel on Climate Change* (Cambridge and New York: Cambridge University Press, 2007).

***Sean Smukler** (sean.smukler@gmail.com) is a postdoctoral research scientist in the Tropical Agriculture and Rural Environment Program of the Earth Institute at Columbia University and **Cheryl Palm** (cpalm@ei.columbia.edu) is a senior research scientist in the Tropical Agriculture and Rural Environment Program, where she is also the science director of the Millennium Villages Project.*

International Food Policy Research Institute
2033 K Street, NW • Washington, DC 20006-1002 • USA
Phone: +1-202-862-5600 • Skype: ifprihomeoffice • Fax: +1-202-467-4439 • Email: ifpri@cgiar.org
IFPRI® **www.ifpri.org**

Agriculture and Climate Change: An Agenda for Negotiation in Copenhagen

2020 Vision for Food, Agriculture, and the Environment

Synergies Among Mitigation, Adaptation, and Sustainable Development

Pete Smith

Focus 16 • Brief 9 • May 2009

There is very significant cost-effective greenhouse gas (GHG) mitigation potential in agriculture. The mitigation potential at a range of future carbon prices is similar to the potential in the industry, energy, transport, and forestry sectors. Using economic mitigation potentials from the Intergovernmental Panel on Climate Change's Fourth Assessment Report (IPCC AR4), the yearly mitigation potential in agriculture is estimated to be worth between US$32 billion and US$420 billion at carbon prices between US$20 and US$100 (t CO_2-eq.$^{-1}$).[1] From both a mitigation perspective and an economic perspective, we cannot afford to miss out on this opportunity. But many mitigation options also offer the promise of facilitating adaptation to climate change and contributing to sustainable development more generally.

In this brief, synergies between mitigation, adaptation, and sustainable development are described so that multiple policy goals can be identified when considering how to include agriculture in the climate change negotiations in Copenhagen.

Identifying the best options for GHG mitigation in agriculture

The IPCC AR4 considered approximately 60 GHG mitigation options in agriculture. These can be grouped into several broad categories including cropland management (such as agronomy, nutrient management, tillage and residue management, and better use of organic manures), grazing land management, agroforestry and set aside, the restoration of cultivated organic soils (such as peats), the restoration of degraded lands, livestock management, manure management, and rice management.

Some of these options work better than others in different regions, so a "one-size-fits-all" recommendation cannot be made. For this reason, the mitigation strategies need to be assessed on a region-by-region basis. The most favorable mitigation options also confer increased resilience and enhanced adaptation to future climate change and/or support sustainable development.

Relationship between mitigation and adaptation in agriculture

Climate change mitigation, impacts, and adaptation will happen simultaneously and interactions will occur. Mitigation-driven actions in agriculture could have either (a) positive adaptation consequences (such as carbon sequestration projects with positive drought preparedness aspects) or (b) negative adaptation consequences (for example, if heavy dependence on biomass energy increases the sensitivity of energy supply to climatic extremes).

Adaptation-driven actions also may have both (a) positive mitigation consequences (as when residue returned to fields to improve water-holding capacity also sequesters carbon) or (b) negative mitigation consequences (for example, an increased use of nitrogen fertilizer to overcome falling yield that leads to increased nitrous oxide emissions).

In many cases, actions will be taken for reasons that have nothing to do with either mitigation or adaptation—for example, actions taken toward enhancing soil fertility or food security. But these events may still have considerable consequences for mitigation, adaptation, or both, as seen when deforestation for agriculture or other purposes results in the loss of both carbon and ecosystems as well as the resilience of local populations. In terms of mitigation, the accumulation rates for sequestered carbon, the growth rates for bioenergy feedstocks, and the size of livestock herds are all variables affected by climate change. Depending upon the climatic impact, there are likely to be shifts in, among other things, plant and tree growth, microbial decomposition of soil carbon, and livestock growth. All of these factors will alter mitigation potential, some positively and some negatively. For example, lower livestock growth rates could increase herd size and, consequently, emissions from manure and enteric fermentation, while increased microbial decomposition under higher temperatures will lower soil carbon sequestration potential. Interactions also occur with adaptation. Crop mix and changes in both land usage and irrigation are all potential adaptation strategies to warmer climates. All would alter mitigation potential.

Mitigation measures that also enhance adaptation

Nearly 90 percent of the mitigation potential in agriculture lies in reducing soil carbon dioxide emissions (by restoring cultivated organic soils, for example) or in sequestering carbon dioxide in the soil organic matter of mineral soils. It has long been known that increasing soil organic matter content improves soil fertility, nutrient supply, soil structure, water-holding capacity, and a host of other vital soil functions. These functions increase the resilience of the soil under threat from future climate change. Soil carbon sequestration then is one of the clearest examples of a mitigation measure that also protects against changes in climate and enhances adaptation and the sustainability of crop production. Other examples include the application of animal manure to soils, which reduces fertilizer use and also improves soil structure and water-holding capacity; the reduction of tillage intensity with improved residue management, which can increase soil carbon while retaining soil moisture; and the restoration of degraded lands, which can sequester carbon and also enhance livelihoods and the resilience of the soils for sustaining agriculture under a changing climate.

Some management changes that are made mainly for the purposes of adaptation—to make agriculture more resilient to climate change—also increase carbon sequestration and so enhance mitigation. For example, conservation tillage increases soil water retention in the face of drought while also sequestering carbon below ground. Small-scale irrigation facilities not only conserve water to cope with greater variability, but also to increase crop productivity and soil carbon stocks. Agroforestry systems increase above- and below-ground carbon storage while also increasing water storage below ground, even in the face of extreme climate events. Properly managed rangelands can cope better with drought and sequester significant amounts of carbon. Project- and program-based funding schemes that support adaptation should also be able to draw on mitigation resources.

[1] The value (in $US) of one ton of carbon dioxide in future carbon markets.

Table 1

ACTIVITY CATEGORY	SUSTAINABLE DEVELOPMENT		
	Social	Economic	Environmental
Agroforestry	Uncertain	Uncertain	Positive
Tillage/residue management[1]	Uncertain	Uncertain	Positive
Nutrient management	Uncertain	Overall efficient use of nutrients will yield cost reduction and productivity improvement	Positive
Water management[2]	Positive	Positive (even if the farmers are supposed to pay for water!)	Positive
Livestock management: breeding improved systems	Uncertain to negative as these practices may not be acceptable due to prevailing cultural practices, especially in developing and underdeveloped society	n/d	n/d
Grazing land management[3]	Positive	Positive	Positive
Increase C storage in agricultural products	Positive	Positive	Positive

[1] Improves fertility of the land. [2] All efficiency improvements are positive for sustainability goals. [3] Positive.

Mitigation and sustainable development

There are various potential impacts of agricultural GHG mitigation on sustainable development. Table 1 evaluates the impact of selected mitigation activities in the agriculture sector on the pillars of sustainable development, namely the social, economic and environmental factors. Table 1 suggests the likely impact, but the exact magnitude will depend upon the scale and intensity of the mitigation measures and where they are undertaken.

Agriculture contributes more than half of the world's emissions of CH4 and N2O; and nutrient, water, and tillage management can help to mitigate these GHGs, especially in rice crops. By careful drainage and effective institutional support, methane emissions and irrigation costs for farmers can be reduced, thereby improving farmer incomes. An appropriate mix of rice cultivation with livestock—known as integrated annual crop-animal systems and traditionally found in West Africa, India, Indonesia, and Vietnam—can increase net income, improve cultivated agro-ecosystems, and enhance human well-being. Such combinations of livestock and cropping, especially for rice, can improve income generation, even in semi-arid and arid areas of the world.

In agriculture in general, groundwater quality may be enhanced and the loss of biodiversity slowed by careful use of farmyard manure and more targeted use of pesticides. The impact of this mitigation measure on social and economic aspects of agriculture, however, remains uncertain. Better nutrient management can improve environmental sustainability.

Pasture improvement by the control of overgrazing favorably impacts livestock productivity (creating greater income from the same number of livestock) and slows or halts soil loss and desertification, thereby providing other environmental benefits. It also provides social security to the poorest people during extreme events such as drought and other crises, especially in Sub-Saharan Africa.

Changes in land cover and tillage management could promote both mitigation and adaptation. A mix of horticulture crops with optimal crop rotations would promote carbon sequestration and could also improve agro-ecosystem function. Societal well-being would also be enhanced through provisioning of water and enhanced productivity. While the environmental benefits of tillage and residue management are clear, other impacts are less certain.

The impacts on sustainable development goals of other mitigation measures listed in Table 1 are context- and location-specific. Appropriate adoption of mitigation measures is likely to help achieve environmental goals, but farmers may incur additional costs, thereby reducing their returns and their income. This trade-off would be most visible in the short term, but, in the long term, synergy among the constituents of sustainable development would emerge through improved natural capital. Trade-offs between economic and environmental aspects of sustainable development might become less important if the environmental gains were better acknowledged, quantified, and incorporated in the decision-making framework.

Suggested negotiating outcomes:

Many mechanisms can be envisaged for rewarding synergies among mitigation, adaptation, and sustainable development, such as giving mitigation credits for projects that also have adaptation potential, giving preference to mitigation projects that also have significant adaptation benefits, or using an adaptation fund. Alternatively, markets could be used to reward synergies, for example by the use of "premium" carbon credits, either as part of future voluntary or compliance markets. Activities that have been shown to confer additional adaptive capacity or enhance sustainable development goals in addition to providing a GHG or carbon benefit would be assigned a higher value in such a credit system than activities that provide only a GHG benefit. ■

For Further Reading: P. Smith et al., "Agriculture," in B. Metz et al., eds., *Climate Change 2007: Mitigation, Contribution of Working Group III to the Fourth Assessment Report of the Intergovernmental Panel on Climate Change* (Cambridge, U.K., and New York: Cambridge University Press, 2007); P. Smith et al., "Policy and Technological Constraints to Implementation of Greenhouse Gas Mitigation Options in Agriculture," *Agriculture, Ecosystems & Environment* 118: 6–28 (2007); P. Smith et al., "Greenhouse Gas Mitigation in Agriculture," *Philosophical Transactions of the Royal Society B* 363: 789–813 (2008).

***Pete Smith** (pete.smith@abdn.ac.uk) is Royal Society-Wolfson Professor of Soils & Global Change at the Institute of Biological and Environmental Sciences, School of Biological Sciences, University of Aberdeen, Scotland, UK.*

International Food Policy Research Institute
2033 K Street, NW • Washington, DC 20006-1002 • USA
Phone: +1-202-862-5600 • Skype: ifprihomeoffice • Fax: +1-202-467-4439 • Email: ifpri@cgiar.org
IFPRI® **www.ifpri.org**

Agriculture and Climate Change: An Agenda for Negotiation in Copenhagen

The Importance of Property Rights in Climate Change Mitigation

Helen Markelova and Ruth Meinzen-Dick

2020 Vision for Food, Agriculture, and the Environment

Focus 16 • Brief 10 • May 2009

Even with abundant evidence of the urgent need for action on climate change mitigation, there are still those who consider mitigation strategies a burden. In the agricultural sector, climate change mitigation calls for changing some agricultural and resource management practices and technologies and often requires additional investment. However, there is an opportunity in agriculture for net benefit streams from a variety of zero- or low-cost mitigation opportunities ranging from agroforestry practices and restoration of degraded soils to zero-till and other land-management practices. Momentum has been generated to incorporate agriculture into carbon markets, potentially allowing smallholder farmers to access benefit streams from such transactions. However, who will receive the benefits from mitigation funds by, for example, increasing carbon stocks or reducing greenhouse gas (GHG) emissions from land, will depend on the way different types of property rights are defined and dealt with in the upcoming climate change negotiations in Copenhagen.

In many areas of the world, land tenure arrangements are complex. For example, in Africa, more than 90 percent of the land is formally claimed as state land, although millions of farming and pastoralist households use various customary and informal arrangements to access the land and other resources. Millions of hectares of forest and pastoral land in Asia and Latin America are similarly listed as state land, although used by communities, especially those of indigenous people or other marginalized ethnic groups. Often the same area may be under co-existing informal tenure systems, most of which are not recognized by formal land laws, but are instead accepted and enforced by the communities. Even where property rights are vested in a formal legal system with strong enforcement procedures, climate change mitigation measures raise new issues of who owns incremental carbon stocks and who should receive compensation for reductions in GHG emissions.

Property rights are complex

The concept of property rights goes beyond formal land ownership. Land users can have any one or a combination of a "bundle" of rights to the land, including access, withdrawal, management, exclusion, and alienation. Rights to land may be separate from rights to trees or other resources. Rights may belong to individuals, groups, or whole communities. They may also come from various sources, ranging from statutory and customary laws and religious practices to international treaties. Limiting carbon sequestration payments to those with formal land titles and individual rights disregards other important sources of claims.

Implication of property rights for mitigation activities

Access to land or other resources may not be enough for smallholders in developing countries to participate in and benefit from climate change mitigation strategies. Tenure security for smallholders is important if they are to take full advantage of schemes such as carbon sequestration payments. It will enhance the welfare impacts of carbon sequestration projects as well as protect the poor and vulnerable from the loss of livelihood sources.

Current funding mechanisms for GHG mitigation through afforestation and agriculture focus only on formal land owners and do not recognize or take into account the complexity of existing tenure arrangements, especially those of millions of people in developing countries. For example, in situations where smallholders do not have a *de jure* right to the land that they use for crop cultivation or livestock rearing, they may not be able to participate in and benefit from afforestation projects outlined under the Clean Development Mechanism (CDM) framework. There are, however, ways in which people without formal tenure rights can be included in mitigation schemes and compensated for the mitigation services they provide. For example, Costa Rican law allows the use of public and private funds to pay landowners without a formal title, promoting several successful payments for environmental service programs such as FONAFIFO (Fondo Nacional de Financiamiento Forestal). Also, in some parts of India, groups of women are compensated for planting trees on local common property. Approaches that similarly recognize and reward those who undertake mitigation activities on customary or common property should be considered for incorporation in future international agreements.

Most low-income households in developing countries live in areas with GHG mitigation potential. Research in natural resource management has found that smallholders will invest in new agricultural technologies (for example, conservation tillage) or practices with long-term benefits (for example, tree planting) only if they have secure rights to the resources. In fact, institutions for secure rights are a precondition for a well-functioning system of payments for environmental services such as carbon sequestration, and they are crucial for the long-term effectiveness of mitigation strategies. If environmental benefits from such transactions are expected in the future, secure property rights are even more important as an incentive for long-term investment in conservation practices, which are essential for both mitigation and adaptation. In some cases, providing *de jure* recognition of customary property rights can both strengthen tenure security and provide incentives for participation in carbon sequestration programs.

Security of tenure and access is needed not just for individuals, but also for communities. In cases where communities hold joint property rights (formal or informal) or, at least, share use and management rights to land, they can act as collective providers of carbon sequestration. Even if a group does not have a formal title to land, as is the case in some pastoralist communities, they can be a valuable supplier of carbon sequestration by, for example, adopting silvopastoral practices that include planting trees and shrubs on pastures to use for fodder and fencing. But many of the current systems risk not only bypassing those without *de jure* land titles; by increasing the value of land, payment systems may even create pressure to alienate such land from those who have been using it under customary or common property rights, thereby leading to the takeover of the land by either the state or large-scale private interest groups. Issues with recent efforts to expand biofuel production in developing countries illustrate the potential problems of paying for carbon sequestration. Large tracts of land allocated by African governments for commercial biofuel cultivation were common property or under customary tenure, but such land is often reallocated without attention to prior usage, which typically results in lost access to resources for long-term land users. For example, in the Bualeba Reserve in Uganda, plans for commercial plantations to generate carbon offset payments under a project threaten to evict local people without formal titles who use the land for farming, grazing, and fishing. In South Africa, the government of Eastern Cape plans to fence off for biofuel production 500,000 hectares of communal land in the Transkei region, which is currently being used for communal grazing and vegetable gardens.

However, there are also examples of how those without a formal land title can be a part of the expanding biofuel markets. The Kavango Biofuel Project in Namibia is a collaboration between local farmers and a Namibian company to grow *jatropha* on communal lands. In return for replacing their maize and millet cultivation with *jatropha*, farmers receive capital costs, food, and cash. Those community members without access to land can participate in other jobs made available through the project.

Suggested negotiating outcome: Include resource users with traditional rights in mitigation funding programs

As mitigation markets grow, there is a danger that poor people with insecure property rights will be excluded. Therefore, mitigation policies should include mechanisms to engage those with legally insecure tenure and to ensure that their rights and livelihoods are not threatened.

Agriculture-based mitigation responses should be designed to include not just the *de jure* owners, but also the users and managers of natural resources with customary rights. Payment recipients should include the users of land who make the investment in mitigation. It is important to consider not only those with individual rights, but those with communal rights as well. The example of the Kavango Biofuel Project illustrates how well-designed mitigation programs can have positive welfare effects for all resource users, regardless of ownership status.

Suggested negotiating outcome: Use mitigation programs to improve tenure security

Millions of hectares of forests, drylands, and other agricultural lands could provide important environmental services for mitigation, but they are held under insecure tenure. In such cases, providing tenure security could be used as an incentive or a reward for participation, either in addition to or instead of monetary compensation. For example, in Indonesia new social forestry agreements increase security of tenure for poor upland farmers in exchange for their commitment to land-management methods that would incorporate agroforestry and land- and water-conservation practices. Such arrangements to secure property rights as part of mitigation programs should be formulated as an early part of the program. Providing stronger land rights as part of the program instead of requiring all participants to have formal title would allow many smallholders to participate and provide incentives that contribute to long-term effectiveness of the programs.

Suggested negotiating outcome: Include pro-poor governance safeguards in mitigation funding mechanisms

Since land is increasingly seen as a target for mitigation strategies, land values are increasing, which presents additional challenges to smallholders without secure tenure. If mitigation programs are poorly structured, then poor smallholders could see their resource rights undermined. Robust safeguards are necessary to protect the poor in the face of growing demands for their land. To ensure that the poor are aware of and educated about their rights to resources, these safeguards must include clear procedures and standards for local consultation, mechanisms for appeals and arbitration, and procedures for informed consent. For example, in Mozambique, the government introduced legislation that requires investors to consult with local communities holding rights to land before undertaking any major commercial enterprise, such as biofuel production.

Regardless of ownership status, payments for carbon sequestration inherently involve a change in property rights over land. By entering those arrangements, the landholder gives up certain management rights, and the payer acquires a partial interest in the land. Therefore, all service providers should have an adequate say in shaping these arrangements. The review of mitigation financing schemes should involve a critical assessment of who holds not only the statutory and customary use rights to resources, but also the decisionmaking rights over those resources. To facilitate such assessment, land information should be improved by updating inventories of land occupation and carefully mapping the bundle of rights that users have. This will contribute to both the effectiveness of carbon sequestration payment programs and the welfare of poor people who manage lands that contribute to climate change mitigation. ■

For Further Reading: IIED and NRI, *Climate Change, Bioenergy, and Land Tenure* (Rome: Food and Agriculture Organization of the United Nations, 2008); B. Swallow and R. Meinzen-Dick, "Payment for Environmental Services: Interactions with Property Rights and Collective Action," in V. Beckmann and M. Padmanabhan, eds., *Institutions and Sustainability* (Dordrecht, The Netherlands: Springer, 2009); S. Pagiola, A. Arcenas, and G. Platais, "Can Payments for Environmental Services Reduce Poverty? An Exploration of the Issues and Evidence to Date from Latin America," *World Development* 33, no. 2 (2005): 237-253; FAO, "Information Sheet on Climate Change and Biofuels," for the High-Level Conference on World Food Security: The Challenges of Climate Change and Bioenergy (Rome: Food and Agriculture Organization of the United Nations, 2008); R. Meinzen-Dick, P. Kameri-Mbote, and H. Markelova, *Property Rights for Poverty Reduction*, 2020 Focus Brief on the World's Poor and Hungry People (Washington, D.C.: International Food Policy Research Institute, 2007).

Helen Markelova *(h.markelova@cgiar.org) is a research analyst and* ***Ruth Meinzen-Dick*** *(r.meinzen-dick@cgiar.org) is a senior research fellow in the Environment and Production Technology Division at the International Food Policy Research Institute.*

International Food Policy Research Institute
2033 K Street, NW • Washington, DC 20006-1002 • USA
Phone: +1-202-862-5600 • Skype: ifprihomeoffice • Fax: +1-202-467-4439 • Email: ifpri@cgiar.org
IFPRI® **www.ifpri.org**

Agriculture and Climate Change: An Agenda for Negotiation in Copenhagen

The Important Role of Extension Systems

Kristin E. Davis

Focus 16 • Brief 11 • May 2009

Climate change will certainly affect agriculture, but agriculture can also be harnessed to mitigate greenhouse gas (GHG) emissions. A key element in supporting agriculture's role is information. The costs of adapting agriculture to climate change can be large and the methods not always well known. Mitigation efforts will require information, education, and technology transfer. Agricultural extension and advisory services, both public and private, thus have a major role to play in providing farmers with information, technologies, and education on how to cope with climate change and ways to contribute to GHG mitigation. This support is especially important for resource-scarce smallholders, who contribute little to climate change and yet will be among the most affected. Support from extension for farmers in dealing with climate change should focus on two areas: adaptation and mitigation, explained below. But first, it is important to define extension.

What is extension?

Extension programs were originally conceived as a service to "extend" research-based knowledge to the rural sector in order to improve the lives of farmers. Extension thus included components of technology transfer, broader rural development goals, management skills, and nonformal education. The traditional view of extension in developing countries was very much focused on increasing production, improving yields, training farmers, and transferring technology. Today's understanding of extension goes beyond technology transfer to facilitation, beyond training to learning, and includes helping farmers form groups, deal with marketing issues, and partner with a broad range of service providers and other agencies. Agricultural extension can thus be defined as the entire set of organizations that support people engaged in agricultural production and facilitate their efforts to solve problems; link to markets and other players in the agricultural value chain; and obtain information, skills, and technologies to improve their livelihoods.

How can extension help with adaptation and mitigation?

There are several ways that extension systems can help farmers deal with climate change. These include adaptation and contingency measures for what cannot be prevented. Extension can help farmers prepare for greater climate variability and uncertainty, create contingency measures to deal with exponentially increasing risk, and alleviate the consequences of climate change by providing advice on how to deal with droughts, floods, and so forth.

Extension can also help with mitigation of climate change. This assistance may include providing links to new markets (especially carbon), information about new regulatory structures, and new government priorities and policies.

Discussed below are three ways in which extension can help with adaptation and mitigation: technologies and management information; capacity development; and facilitating, brokering, and implementing policies and programs.

Technologies and management information

Extension traditionally has played a role in providing information and promoting new technologies or new ways of managing crops and farms. Extension also links farmers to researchers and other actors in the innovation system. Farmers, extension agents, and researchers must work together on farmers' fields to prioritize, test, and promote new crop varieties and management techniques. While extension must now go beyond such methods, there is still a need for simple technology transfer in order to increase resilience to climate change and mitigate GHG emissions

Today's farmers will need to be able to quickly respond to climate change and adeptly manage risk. This will be especially challenging for extension in terms of knowledge and information systems. Farmers need to have access to this kind of information—be it climatic information, forecasts, adaptive technology innovations, or markets—through extension and information systems.

Extension agents can introduce locally appropriate technologies and management techniques that enable farmers to adapt to climate change by, for example, developing and disseminating local cultivars of drought-resistant crop varieties with information about the crops' advantages and disadvantages. Additionally, extension staff can share with farmers their knowledge of cropping and management systems that are resilient to changing climate conditions such as agroforestry, intercropping, sequential cropping, and no-till agriculture. Some of these practices have the added advantage of improved natural resource management. Tree planting can also help to improve soil, prevent soil erosion, and increase biodiversity. It is important to provide farmers with information about how the various options will potentially increase income and yields, protect household food security, improve soils, enhance sustainability, and generally help to alleviate the effects of climate change. At the same time, extension staff can play an important role in transferring indigenous technical knowledge to help farmers worldwide.

A core challenge for extension in the future is to shift from providing "packages" of technological and management advice to, instead, supporting farmers with the skills they need to choose the best option to deal with the climate uncertainty and variability and to make informed decisions about if and how to engage in new markets for carbon emissions. Some farmers will also need access to new technologies and management options in those areas where climate change renders their current farming systems inviable.

Capacity development

One of extension's major activities over time has been adult and nonformal education. This role continues today and is even more important in light of climate change. In addition, extension is also responsible for providing information using techniques ranging from flyers and radio messages to field demonstrations. Recent innovative extension activities include the adult education and experiential learning approaches utilized in farmer field schools, an extension and education approach already working with farmers on issues of climate change. Climate Field Schools (CFSs) have been established in West Java, Indonesia, to deal with climate change in agriculture. Another example is a multimedia campaign planned by True Nature Kenya and the World Agroforestry Centre that will show films and offer educational follow-up by extension agents to publicize grassroots solutions to the problems of climate change.

Climate change will initiate extreme events like sudden onset disasters and new vectors of human and livestock diseases. Evidence is emerging that the biggest impacts will be in the form of small droughts, floods, and other events that cause severe hardship but do not attract the attention of the international community. The capacity of farmers to cope with such different forms of risk will become ever more crucial, and extension efforts must pay special attention to educating farmers about their options to enhance resilience and response capacity (see also the brief on extreme events). There is a need for capacities to engage new sets of actors, including humanitarian agencies. Education must thus move beyond technical training to enhance farmers' abilities for planning, problem solving, critical thinking, prioritizing, negotiating, building consensus and leadership skills, working with multiple stakeholders, and, finally, being proactive.

Capacity development is important within extension as well. Extension agents have traditionally been trained only in technical expertise and often lack "soft" skills such as communication, development of farmer groups, systems thinking, knowledge management, and networking. To improve outcomes in rural development, farmers and extension agents need new skills that will require agricultural education and extension curriculums to include valuing and understanding the knowledge and experiences of rural people and co-learning (that is, farmers and extension agents learning together rather than extension agents training farmers in a one-way information transfer). There are many different ways to inform and educate farmers about adaptation options. Climate change adaptation funding should focus on extension systems and programs that incorporate a good understanding of what practices and skills are needed to best promote activities that help in the climate change effort and on increasing the capacity of extension agents and farmers, where needed.

Facilitating, brokering, and implementing policies and programs

Another role of extension, which will be critical for climate change issues, is that of acting as an honest broker, bringing together different actors within the rural sector. Traditionally this has meant linking farmers to transport agents, markets, and inputs suppliers, among others.

With climate change, it will be increasingly important for the extension system to link farmers and other people in rural communities directly with voluntary and regulated carbon markets, private and public institutions that disseminate mitigation technologies, and funding programs for adaptation investments. Increased access to meteorological information will be imperative. Extension also has an enormous challenge in bringing together farmers' concerns and those of other actors as they address both climatic and market uncertainties together. Extension has the chance to make a significant contribution to overcoming this gap through enhanced farmer decisionmaking.

Extension agents may also play a role not only in brokering, but also in assisting farmers in implementing policies and programs that deal with climate change mitigation. For instance, regarding carbon credits, extension agents could be employed to educate farmers in their area; assist in forming community groups; link farmers to governmental, nongovernmental, and private organizations at the national and international levels; and perhaps assist with proposal preparation or negotiations with other players.

Why extension rather than another institution for climate change?

Gathering information is expensive. Extension has proven itself to be a cost-effective means of bringing about greater economic returns for farmers with significant and positive effects on knowledge, adoption, and productivity. Studies of extension productivity report rates of return from 13 to 500 percent. A recent study demonstrated that receiving at least one extension visit in Ethiopia reduced smallholders' likelihood of being poor by 10 percent and increased consumption growth by 7 percent. Extension is thus a cost-effective tool that can play an important role in dealing with climate change while at the same time helping to increase productivity and reduce poverty.

Suggested negotiating outcomes:

Extension has a major role to play in helping farmers adapt to and mitigate climate change. To capture this potential role, adaptation and mitigation funds could be used to support extension efforts that deliver new technologies, information, and education about increasing carbon sequestration and reducing GHG emissions. Traditionally extension has worked to promote new technologies and management techniques, educate farmers, and act as a facilitator or broker for rural communities. Now, too, extension can help link practice in the field to new policies regarding climate change. All of these roles can be exploited in a cost-effective way to help resource-poor smallholders deal with the issues of climate change that will so radically affect their livelihoods. Perhaps the most important purpose for extension today is to bring about the empowerment of farmers, so that their voices can be heard and they can play a major role in deciding how they will mitigate and adapt to climate change. ■

For Further Reading: D. Birkhaeuser et al., "The Economic Impact of Agricultural Extension," ***Economic Development and Cultural Change*** **39, no. 3 (1991): 607–640; S. Dercon et al.,** ***The Impact of Agricultural Extension and Roads on Poverty and Consumption Growth in Fifteen Ethiopian Villages,*** **Discussion Paper No. 840 (Washington, D.C.: International Food Policy Research Institute, 2008).**

Kristin E. Davis *(k.davis@cgiar.org) is a research fellow in the Knowledge, Capacity, and Innovation Division at the International Food Policy Research Institute (IFPRI).*

International Food Policy Research Institute
2033 K Street, NW • Washington, DC 20006-1002 • USA
Phone: +1-202-862-5600 • Skype: ifprihomeoffice • Fax: +1-202-467-4439 • Email: ifpri@cgiar.org
IFPRI® **www.ifpri.org**

2020 VISION FOR FOOD, AGRICULTURE, AND THE ENVIRONMENT

Adaptation to Climate Change: Household Impacts and Institutional Responses

FUTOSHI YAMAUCHI AND AGNES QUISUMBING

FOCUS 16 • BRIEF 12 • MAY 2009

Climate change will bring with it increased frequency of two types of natural disasters that affect agriculture and rural households: droughts and floods. It will also alter rainfall patterns, thereby changing farming practices, household behavior, and welfare.

Households all over the world use a variety of formal and informal mechanisms to manage risk and cope with unexpected events that negatively affect incomes, assets, or well-being. These mechanisms include both preparation for and responses to natural disasters. In low-income settings, where formal insurance and government supports are limited, households tend to rely on informal coping strategies, such as transfers from friends and neighbors, remittances, or investments in a diverse range of assets, from livestock to human capital. When disaster-related shock affects only a few households at a time, informal mechanisms can be quite effective in dealing with the situation. However, if the shock affects large areas simultaneously, small-scale coping mechanisms become ineffective.

Research on several climate-related national disasters—the 1998 floods in Bangladesh, the 2001 drought in Ethiopia, and the 2001–02 failed maize harvest in Malawi—suggests that the upcoming negotiations in Copenhagen need to explicitly define, support, and expand policies that protect vulnerable populations from the expected increase in climate-change related weather events.

Household responses: Ex ante and ex post

People adopt different response strategies based on the scale of a perceived hazard. Because individuals cannot distribute risk equally among other individuals and households, they need to reallocate resources accordingly, either by accumulating physical or human capital assets over time or by entering into contracts that are valid only if certain outcomes occur.

Households in the three disaster-prone regions studied had similar responses to the effects of a natural disaster. Relatively low disaster probabilities do not sufficiently motivate households to invest in assets to diversify against risk. However, if the perceived threat of disaster is sufficiently high, they will allocate more of their resources to human capital and livestock, which are both relatively mobile. For example, in Bangladesh and Malawi, households increased human capital relative to land, while, in Ethiopia and Malawi, households held more livestock relative to land. These differences suggest that beneficial ex ante actions depend on market returns to these assets as well as the effectiveness of ex post institutional responses to previous disasters. The importance of ex ante actions will increase as natural disasters occur more frequently.

Households with more schooling are also better prepared in both the short and long term. Human capital is less affected by natural disasters than physical capital; it is portable and remunerable in different locations and obtains more stable returns. Additionally, better health and nutritional status of children raises their survival probability and resilience to disasters. Taller children are less likely to become sick even in unsanitary post-disaster environments. In Bangladesh, for example, taller children were less affected by the adverse effects of the 1998 floods and made up missed schooling much sooner than shorter children. In Ethiopia and Malawi, the exposure to highly frequent droughts reduced schooling for some, but with more negative impacts on shorter children.

The significance of ex ante actions and preparedness also depends on the effectiveness of public assistance, both before and after a disaster. In Ethiopia, public assistance after disasters played a more important role than ex ante actions in mitigating the impact of the shocks on child schooling. In contrast, Malawi relied on private ex ante actions, since public aid was largely insignificant. Both ex ante and ex post actions were important in Bangladesh.

Institutional responses: Targeting and effectiveness

IFPRI conducted studies of governmental emergency assistance provided in three climate-related natural disasters. Evaluations were based on detailed household surveys completed between eighteen months and five years after the disaster and, in most cases, several months after emergency food aid disbursements had stopped.

Bangladesh

Following the 1998 floods, two existing relief programs provided the bulk of food assistance. Gratuitous Relief (GR) provided free food targeted both across and within localities, with community-level decisionmakers allocating relief directly to the most affected households. The Vulnerable Group Feeding (VGF) program also used community targeting, employing criteria such as assets, income, occupation, and demographics (for example, female-headed households). GR was mobilized immediately after the flood, while VGF was implemented months later and was broader in scope. GR community relief committees appear to have channeled relief more effectively to flood victims, while VGF directed food more to the poor rather than to those severely affected by the flood. GR functioned better as a disaster-relief mechanism, suggesting that community-level decisionmakers are better equipped to decide who should receive emergency assistance.

Ethiopia

After a severe drought in 2002, two types of emergency assistance were made available in the most severely affected areas: Food-for-Work (FFW) and targeted free food distribution or gratuitous relief (GR). Ethiopia's National Food Aid Targeting Guidelines gives local communities responsibility for creating criteria for the allocation of drought relief. For public work provided under FFW or Employment Generation Schemes (EGS), locally set targeting criteria seem to have been poorly understood. Instead of targeting resources to those most severely affected by the drought, EGS reached out to households with able-bodied individuals who were willing to work and set the wage in such a way that the richest households were only slightly less likely to participate. Targeting criteria were better understood in the GR program.

Malawi

When Malawi's maize harvests failed in 2001 and 2002 after a severe drought, the initial response to the food crisis was delayed due to poor information, trade and transport bottlenecks, and a general lack of institutional preparedness in dealing with large-scale emergencies. Food assistance arrived late, although it was quickly scaled up. Most of the aid was administered by nongovernmental organizations in partnership with district- and village-level institutions, using community-based targeting. The largest program, General Food Distribution, targeted the "poorest of the poor," with a special emphasis on households with orphans and/or malnourished children, families with elderly or ill members, female-headed households, and those who had suffered the most from the drought.

IFPRI research found that when targeting criteria are easily verified—for instance, female-headed households or households with orphans—aid institutions were more likely to abide by the targeting guidelines. When the criteria were not easily identified, as in the case of defining the "poorest of the poor," they were less likely to be implemented.

Community-based targeting worked well in Bangladesh, moderately in Ethiopia, and minimally in Malawi. Three features are of note. First, targeting experience improves performance. Bangladesh, the most successful in reaching targeted populations, also had the most experience with targeting emergency assistance on a unified, national scale. Ethiopia, while responding to droughts over a similar period of time, had a historically decentralized approach, with national coordination being relatively recent. Malawi had not experienced a similar crisis since the 1940s and was ill prepared for a national food emergency. Second, community-based targeting is more effective under tight budgets. For example, in Ethiopia, where communities allocated both public works and free distributions, targeting was better in the latter program, which was more resource-constrained. Finally, more information within communities can improve the effectiveness and consistency of targeting as well as increase residents' trust that relief is being disbursed fairly and rationally.

In addition to mitigating the short-term effects of natural disasters on food consumption, targeted emergency food relief can reduce households' need to sell physical and human assets, thereby having longer term impacts on asset holdings and the overall future well-being of those affected by disaster. It appears that the long-term effects of either Food-for-Work or free food distribution on asset holdings and consumption was limited, although positive impacts were found for some groups of recipients in all three countries. In Bangladesh, GR was effective in protecting the asset levels of the poorest quintile while, in Ethiopia, GR recipients in the two poorest quintiles had higher livestock holdings than nonrecipients. There is some evidence that food aid's role in protecting assets may have persistent effects on food consumption as well. However, aid does not always benefit the poorest of a given population. In Ethiopia, for example, households benefiting the most from EGS participation were in the middle and upper portions of the expenditure distribution.

Suggested negotiating outcomes:

Climate change adaptation measures should include adequate funding and preparation for emergency assistance to respond rapidly to climate-change related disasters. Launching relief efforts as early as possible in an emergency prevents people from using coping mechanisms that harm health or nutritional status (for example, by reducing the number of meals they eat) or compromising their livelihoods (by selling productive assets). Early action requires reliable early warning systems. Also, food aid has greater impact when the schedule of assistance—even if it is brief—is well known and consistent, because this allows recipients to plan consumption and investment.

Relief providers should consider increasing the ration size since food-aid rations, in all cases, amounted to only a small proportion of household consumption. Increasing rations to specific households would mean decreasing the number of aid recipients, therefore targeting would need to be more effective.

Given limited aid resources and the probability that climate change will cause more frequent disasters, existing social protection systems will be increasingly strained. They will need to be redesigned to account for new threats to the livelihoods and well-being of poor households. Thus, funding adaptation that arises from the Copenhagen negotiations should reflect these key findings:

- More education and nutrition ex ante are important. While investment in education and nutrition is part of good development policy in general, it also facilitates adaptation to climate change.
- Ex post targeting by communities, focused on those most severely affected by or least able to cope with disaster, is essential. Easily verifiable indicators of vulnerable groups—including female-headed households, orphaned children, and so forth—will facilitate local targeting. Funding will have to be devolved downward to the extent possible and communities empowered to make decisions on emergency assistance allocations.
- Countries with more experience in managing natural disasters perform better over time, so intergovernmental mechanisms should enable countries to learn from one another's experiences. ■

For Further Reading: D. O. Gilligan et al., *Assessing the Effectiveness of Community-Based Targeting of Emergency Food Aid in Bangladesh, Ethiopia, and Malawi,* IFPRI-WFP Research Action Brief (Washington, D.C., and Rome: International Food Policy Research Institute and World Food Programme, 2005); D. O. Gilligan et al., *Assessing the Longer-Term Impact of Emergency Food Aid in Bangladesh, Ethiopia, and Malawi,* IFPRI-WFP Research Action Brief (Washington, D.C., and Rome: International Food Policy Research Institute and World Food Programme, 2005); F. Yamauchi, Y. Yohannes, and A. Quisumbing, *Risks, Ex-ante Actions and Public Assistance: Impacts of Natural Disasters on Child Schooling in Bangladesh, Ethiopia, and Malawi,* Policy Research Working Paper No. 4909 (Washington, D.C.: World Bank, 2009).

***Futoshi Yamauchi** (f.yamauchi@cgiar.org) is a research fellow and **Agnes Quisumbing** (a.quisumbing@cgiar.org) is a senior research fellow in the Poverty, Health, and Nutrition Division at the International Food Policy Research Institute.*

International Food Policy Research Institute
2033 K Street, NW • Washington, DC 20006-1002 • USA
Phone: +1-202-862-5600 • Skype: ifprihomeoffice • Fax: +1-202-467-4439 • Email: ifpri@cgiar.org
IFPRI® **www.ifpri.org**

Agriculture and Climate Change: An Agenda for Negotiation in Copenhagen

The Constructive Role of International Trade

Franz Fischler

2020 Vision for Food, Agriculture, and the Environment

Focus 16 • Brief 13 • May 2009

An open and flexible global trading environment plays a constructive role in both climate change mitigation and adaptation. A new international climate change regime and global trade rules should ideally be mutually reinforceable.

The importance of an open trading system

Trade of food and agricultural products will be crucial in order to offset climate change-induced reduction of production in certain regions and to compensate for periodic shortages due to more frequent droughts and floods. An open trade system can also help address volatility of global agricultural prices, which is likely to be exacerbated by the impacts of climate change.[1]

Agricultural expansion is a key driver of deforestation. The global community is therefore well advised to focus on ways to increase agricultural productivity for both crops and livestock on existing arable land. Not only will this lead to improved food security, but it is also an important indirect form of greenhouse gas (GHG) mitigation. Easy and reliable access to technologies will be crucial in moving toward a low-carbon global economy. This, in turn, will be facilitated by an open trading system, especially for yield-increasing inputs into agricultural production.

It is not desirable to increase yields without concern for the environment through, for example, a high amount of fertilizer use. Certain regions, such as Africa, however, can easily increase their use of fertilizers, given their current low levels of use. Other regions must adapt to improved fertilizer practices in order to decrease nitrous oxide emissions without sacrificing yield. Improved seeds, including those derived from modern biotechnology, not only increase yields, but also have the potential to help producers grow their crops in drier and hotter conditions and can maximize the nitrogen uptake from fertilizers. Indeed, it is hopeful that new biotech crops with such useful traits will sway those skeptical of the technology and thus help reduce present regulatory trade barriers.

Trade-distorting domestic support to agriculture, which is tied to production, should be reduced—or, even better, eliminated—since it can lead to environmental degradation by overexploiting scarce or fragile natural resources. It is instructive, in this regard, to see how the reforms of the Common Agricultural Policy (CAP) have been accompanied by a steady reduction of GHG emissions from agriculture in the European Union, for example, through a reduction of livestock herds. A decoupling of domestic support from production in Organisation for Economic Co-operation and Development (OECD) countries also contributes to increased investment in the agricultural sectors of developing countries.

The Doha Development Round includes a negotiation on lowering tariffs (and other non-tariff barriers) on environmental goods and services. Some have expressed the view that the opposite principle could also be applied, namely that higher tariffs (or border tax adjustments) could be used to discourage imports of products from countries without climate change regulations, or of limiting demand for products with large carbon footprints. Instead, emphasis should be placed on ensuring greater commitment to mitigation in all countries. A comprehensive carbon trading scheme (or tax) should result in prices that reflect a product's relative carbon intensity and will be more effective than governments resorting to tariff measures at the border.

Recommendations:

- By allowing food products (including embedded water) to move to regions negatively impacted by climate change, trade contributes to adaptation and global food security.
- Trade can help address volatility of global agricultural prices, which is likely to be exacerbated by the impacts of climate change.
- Increasing agricultural productivity on existing arable land, thereby reducing pressure on forests, must be an important climate change mitigation and adaptation strategy. Liberalized trade in goods and services will assist in this process.
- The international community should continue to reduce and eliminate trade-distorting domestic support, as this frees up investment in agricultural sectors in developing countries and leads to more efficient and sustainable global production patterns.
- Tariffs or border-tax adjustments are not good instruments for limiting imports of products from countries without sufficient climate change regulation, specifically because they lead to discrimination and retaliation.

International climate change and trade rules should be coherent

There is a great deal of speculation about possible conflicts between international climate change and international trade rules. It is important for the international community to strive for policy coherence. In theory, there should be no conflicts between these two sets of international rules. The United Nations Framework Convention on Climate Change explicitly states that measures taken to combat climate change should not constitute a means of arbitrary or unjustifiable discrimination or a disguised restriction on international trade. Likewise, World Trade Organization (WTO) ministers have pledged that an open, nondiscriminatory, multilateral trading system and actions that protect the environment and promote sustainable development can and must be mutually supportive.

The WTO's insistence on national treatment and nondiscrimination may well serve as a sufficient bulwark against countries tempted to hide protectionist motives under the guise of climate change. Likewise, the WTO rules may already be sufficiently flexible to adjust to the new climate change rules. Therefore, the question of aligning WTO rules should be tackled after the conclusion of a new international climate change regime. WTO parties should devote their efforts to a rapid conclusion of the Doha Development Round. The economic and developmental benefits from concluding the round will strengthen countries' abilities to address climate change.

WTO members should consider adopting a peace clause, which

[1] A considerable number of export restrictions were put into place in 2007 and 2008 to mitigate against rising food prices, but, actually, they blocked important supply responses and contributed to higher global prices. See S. Mitra and T. Josling, *Agricultural Export Restrictions: Welfare Implications and Trade Disciplines*, IPC Position Paper, www.agritrade.org/GlobalExpRestrictions.html (2009).

exempts certain climate change measures from litigation under the WTO for a short period. The purpose of such a clause would be twofold. It would allow the international community to understand how various countries choose to implement a new international climate change regime. It would also provide a period of time for countries to consider whether specific amendments to WTO rules or to the new climate change rules will be required.

An important area for consideration will be the trade of carbon credits. A "trade and cap" regime is seen by many as the most promising approach for combating climate change. Such regimes can be established at the national or regional level. In the long run, a global carbon market would lead to the most cost effective emissions reductions and importantly reduce the risks of leakage, but, for the foreseeable future, efforts must be placed on how national or regional schemes should be interlinked. As with other trading arrangements, these carbon trading schemes will require sound rules. There are issues such as the scientific definition and measurement of units and their interoperability between different sectors and countries. Key concepts from the General Agreement on Tariffs and Trade (GATT)/WTO sphere, such as non-discrimination and national treatment, may also need to figure in an international carbon trading system in order to ensure a level playing field. Concerns that widely differing national methods for allocating allowances or accepting offsets may impair fair competition also have to be addressed.

The growing literature about the need to clarify WTO rules with regard to climate change also encompasses border tax adjustments; what is and is not allowed under the WTO's exceptions clause, Article XX; and whether or not products can be differentiated from each other based on their production and processing methods. All these topics are highly relevant for the agricultural sector and should be included in a work program for policy coherence between international climate change and trade rules. There are, however, a number of issues specific to food and agriculture, which deserve additional attention:

Given agricultural production's reliance on soil and water, there will be an increasing number of marketing incentives intended to promote environmentally friendly agricultural products. Examples include information on the carbon or water footprint of a product. Producers in developing countries may find it difficult to meet a proliferation of distinct requirements, so care will have to be taken to ensure transparency and greater coherence among private-sector schemes. As far as government standards are concerned, the WTO clearly identifies three international standard-setting bodies in the sanitary and phytosanitary realms whose task it is to arrive at internationally approved food standards. National measures based on these standards are automatically considered to be WTO compliant. The international community could consider establishing a similar environmental or climate change-standard-setting body, which would seek consensus on complex scientific issues, such as how to calculate a lifecycle carbon analysis.

Another important consideration pertains to government support provided to agricultural producers for mitigation measures. Subsidies are already widely used in the production of biofuels and could be used to incentivize carbon sequestration in soils or reduce agricultural emissions in other ways. When public money is used toward that end, WTO rules on subsidies will come into play. In addition to the WTO's antidumping and countervailing rules, the Agreement on Agriculture's rules on subsidies will also need to be examined. WTO rules may be helpful in ensuring that countries do not disguise increased levels of trade-distorting support under the guise of climate change. Alternatively, however, WTO rules may prove to be too restrictive for genuine climate change measures.[2]

Recommendations:

- The Doha Round modalities should not be reopened in order to address climate-change related issues. Countries should seek to conclude the Doha Round now in its own right. The economic and development benefits from concluding the Round will strenghten countries' ability to address climate change.
- WTO members should consider establishing a task force or work program to examine the relationship between existing trade rules and climate change measures. In order to avoid drawn out and difficult disputes in the WTO, members can withhold from litigation when newly agreed international climate change rules are in the process of implementation at the national level.
- An important question which must be addressed is the extent to which international trade rules should also be applied to the international trade of carbon credits, and if not, how a separate set of trade rules for carbon credits is to relate to WTO rules.
- A greater harmonization of both public- and private-sector climate change-related standards should be pursued in order to make these more effective and less trade distorting.
- To avoid potential conflicts in the area of agricultural GHG reduction subsidies, a "health check" for internal domestic support (in other words, an examination of whether the Agreement on Agriculture rules on domestic support need to be revised) may be advisable.
- An open and flexible trading system will be an important factor as the world looks for ways to adapt to climate change as well as to mitigate it by reducing GHG emissions. To avoid conflicts between international climate change and trade rules, the international community needs to ensure coherence. It is, after all, the same countries negotiating in Bonn (and elsewhere under the UNFCCC umbrella) as in Geneva. ■

For Further Information: To examine the interlinkages between climate change, agriculture, and trade, the IPC and the ICTSD have convened a "Platform on Climate Change, Agriculture, and Trade: Promoting Policy Coherence." See www.agritrade.org/events/ClimateChangePlatform_000.html.

---

[2] See T. Josling and D. Blandford, *Greenhouse Gas Reduction Policies and Agriculture: Implications for International Trade Disciplines*, forthcoming from the International Centre for Trade and Sustainable Development (ICTSD)–International Food & Agricultural Trade Policy Council (IPC) Platform on Climate Change, Agriculture, and Trade.

***Franz Fischler** (ffc@franz-fischler-consult.co.at) is the former European Commissioner for Agriculture and a member of the International Food & Agricultural Trade Policy Council (IPC).* *He would like to thank Charlotte Hebebrand, IPC's chief executive, for her contributions to this brief.*

International Food Policy Research Institute
2033 K Street, NW • Washington, DC 20006-1002 • USA
Phone: +1-202-862-5600 • Skype: ifprihomeoffice • Fax: +1-202-467-4439 • Email: ifpri@cgiar.org
IFPRI® **www.ifpri.org**

Made in the USA
Monee, IL
05 November 2026

67379794R00019